Analogue Video Technological evolution plus DIY circuits

Angelo La Spina

All rights reserved. No part of this book may be reproduced in any material form, including photocopying, or storing in any medium by electronic means and whether or not transiently or incidentally to some other use of this publication, without the written permission of the copyright holder except in accordance with the provisions of the Copyright, Designs and Patents Act 1988 or under the terms of a licence issued by the Copyright Licensing Agency Ltd, 90 Tottenham Court Road, London, England W1P 9HE. Applications for the copyright holder's written permission to reproduce any part of this publication should be addressed to the publishers.

The publishers have used their best efforts in ensuring the correctness of the information contained in this book. They do not assume, and hereby disclaim, any liability to any party for any loss or damage caused by errors or omissions in this book, whether such errors or omissions result from negligence, accident or any other cause.

British Library Cataloguing in Publication Data
A catalogue record for this book is available from the British Library

ISBN 978-0-905705-96-5

Illustration 'Video Painter' on cover: Angelo 'Tecchese' La Spina © 2011
Prepress production: Kontinu, Sittard
First published in the United Kingdom 2011
Printed in The Netherlands by Wilco, Amersfoort
© Elektor International Media BV 2011

109025-UK

Contents

Contents 3

Biography 7

Preface 8

Chapter 1: The Universe of Video 12

Chapter 2: Television Standards 15
2.1   History and Prehistory of Television.. 15
2.2   What standard to choose?.. 21
2.3   The waveforms of synchronisms 23
2.3.1   Horizontal & Vertical pulses 24
2.3.2   Interlace 2:1 25
2.3.3   Broad & Equalisation pulses 25
2.3.4   Vertical frequency 26
2.3.5   Aspect ratio 27
2.3.6   Video Bandwidth 28
2.4   Sync Pulse Generators (SPGs) 29
2.4.1   Single chip SPGs 30
2.4.2   Multiple chip SPGs 35
2.4.3   Non-standard SPGs 39
2.5   Sync extraction 40
2.6   A curious television scanning alternative! 44

Chapter 3: Colour Television Standard Systems 46
3.1   The Human Eye 46
3.2   From eye to its electronic representation 47
3.3   NTSC & PAL Encoding 52
3.4   SECAM System 59
3.5   The relative merits and demerits of TV standards 60
3.5.1   NTSC/525-lines pros 61
3.5.2   NTSC/525-lines cons 61
3.5.3   PAL/625-lines pros 61
3.5.4   PAL/625-lines cons 62
3.5.5   SECAM/625-lines pros 62

3.5.6	SECAM/625-lines cons	62
3.6	Other analogue colour television systems	64
3.7	Colour Encoders	65
3.7.1	Digital Inputs Encoder Chips	65
3.7.2	Analogue Input Encoder Chips	68
3.8	NTSC & PAL Colour Decoding	76
3.9	SECAM decoding	79
3.10	Colour Decoder Chips	81
3.10.1	Analogue Decoders	81
3.10.2	SECAM decoders	89
3.10.3	Digital & Pseudo-Digital Decoders	91
3.10.4	Digivision TV chipset	94
3.11	Video conversion	96
3.11.1	From PAL S-Video to CVBS and vice versa	97
3.11.2	From YUV to RGB and vice versa	98
3.12	Some final considerations	99

Chapter 4: Pattern & monoscope generators 101

4.1	The Colour Test Pattern Bars Generator	102
4.2	Patterns for TV set calibration	108
4.3	Multiburst Generator	110
4.4	An automatic RGB Colour Changer	113
4.5	Test Cards & Monoscopes	114
4.6	Software based Test Card Generators	117
4.7	Two final remarks	121

Chapter 5: Television display systems 124

5.1	A little bit of history...	124
5.2	Display Devices	126
5.2.1	The Oscilloscope	127
5.2.2	The Waveform Monitor	130
5.2.3	A LED Oscilloscope	131
5.2.4	The Vectorscope	133
5.2.5	The Baird's 'Radiovision', the first industrial mechanical television set	134
5.2.6	The Cathode Ray Tube (CRT) or Kinescope	135
5.2.7	Other types of CRTs	142

5.2.8	The Field Emission Display (FED)	143
5.2.9	The Surface-conduction Electron-emitter Display (SED)	144
5.2.10	The Liquid Crystal Display (LCD)	144
5.2.11	The Active Matrix Liquid Crystal Display (AMLCD)	145
5.2.12	The HDR-TV Display	146
5.2.13	The Plasma Display Panel (PDP)	146
5.2.14	The Organic Light Emitting Diode (OLED) Display	147
5.2.15	The Thick-Film Dielectric Electroluminescent (TDEL) Display	148
5.2.16	The WOWvx Display	149
5.2.17	The Virtual Retinal Display (VRD)	150
5.2.18	Large screen displays	150
5.2.19	Multiple screen displays	153
5.3	Things to Think About	154

Chapter 6: Video Cameras 156

6.1	Tube-based Camera Devices	156
6.1.1	The first experimental camera tubes	157
6.1.2	Further developments on camera tubes	158
6.1.3	1, 2, 3 or even 4-tube Colour Cameras	161
6.1.4	Pickup tube features	164
6.2	Solid-state Pickup Devices	165
6.2.1	The CCD (Charge Coupled Device)	165
6.2.2	CMOS image sensors	169
6.3	Features common to all Cameras	170
6.4	Some final considerations	172

Chapter 7: Video recorder systems 174

7.1	Video recording chronological history	174
7.2	Video cassette recorder mechanisms	183
7.3	Magnetic playback and recording techniques	187
7.4	Footage transferring	189
7.5	Analogue Anticopy Protection	191
7.5.1	How does the anticopy system work?	192
7.5.2	How to eliminate protection pulses	193
7.5.3	Description of the circuit	193
7.5.4	The output stage & Power Supply	195

7.5.5	Assembling & Soldering	195
7.5.6	Calibration	196
7.5.7	Inside a box	197
7.5.8	Bad thoughts	198
7.6	Some conclusive considerations	198

To be continued **200**

Bibliography **201**

Index **205**

About the Author

Angelo La Spina was born in August 27th, 1961, in Riposto (Italy), a small town between the Ionian Sea and the volcano Etna. He qualified in 1980 in an Italian scientific secondary school with honours. During the same studies he attended a correspondence course about electronics, specifically radio and television technologies, getting the relative certificate in March 1979 with honours too.

This could be considered a starting point for his career in television sector since in 1980 he joined the staff of employees and technicians of one of the first private Italian TV stations, by now closed. There he designed and built by himself some video equipment such as a video mixer and a memory-based colour-overlay logo generator as well as other minor devices. In Italy, that was a period of television pioneering experimentation due also to a legislative lack of TV channel regulation so that hundreds of unregulated private TV stations were born.

Following this revolutionary spirit, in 1982 he left the TV station to submit his video device designs to a local company which marketed them throughout Italy. Following on from this success, other video-compliant units were designed and produced including switchers, cross matrixes, amplifiers, bars generators, Amiga computer synchronizers (AKA *Amiga Genlocks*), colour correctors, etc. Also he advised and maintained dozens of TV stations, some of which were so tiny that their telecasts were limited to a few square kilometres!

In 1991 the boom of private TV stations finished due to the heavy intervention of the Italian Government (which has closed several of them) and he resigned to dedicate himself to the incoming digital and computer technologies.

Meantime from 1982 to 1986 he qualified with honours after attending four courses as a computer operator, solar water heating installer, electrician and electromechanical worker.

Since 1992 he is still working 'jumping' among the jobs of freelance computer operator, NLE video editor, CGI operator, cameraman, photographer, radio & TV repairer, electrician, PC builder, repairer, seller, dealer and consultant, DTP operator, PHP/Mysql/Java/Javascript/HTML/CSS code programmer and web designer, electronic project-designer on demand, gaining so a huge experience in several fields.

In 1995 he rendered a short CGI animated 3D movie named *Duel* which was presented to some competitions (*Premio Immagine 95*, Milan, 4th place; *Bitmovie 96*, Rimini, 6th place; *Creativa 96*, Faenza, 1st place). Years later (2008) the remake of the same movie was chosen by the organizers of the *Clorofilla Film Festival* (Grosseto), a show sponsored by *Legambiente and the Italian Ministry for the Environment and Territory*.

From 2003-2004 he has taught computer science at a state school and later (2007-2008) he has edited and directed a local monthly magazine specialized in Sicilian culture.

In January 2009 he began to write this book for you.

This book is a tribute to over thirty years of his life dedicated to electronics, computer science, video and television technologies.

Enjoy!

Preface

The Civilization of Image

Fig. 1 - John Logie Baird

About the time technology enabled man to transmit sound over long distances, first through telephone wires and later radio waves, man began to cherish the dream of reducing the space using images. Now, that would be predictably difficult because of all the evident difference between the mechanisms of handling sound in contrast to vision.

As in telephony the ear had been replicated by the microphone to translate sound into electrical pulses, so at first the eye seemed to inspire a corresponding technical achievement in television. The difference is that the wonderful mechanism of vision for which the visual perceptions of the retina are routed in the optic nerve and <u>simultaneously</u> processed by the brain, will be almost impossible to imitate perfectly or even surpass.

More specifically, in normal vision, the front lens of the eye focuses the image of the subject on the retina whose surface is formed by numerous photoreceptor cells (rods and cones). These become simultaneously excited in proportion to the luminous intensity of the various points of the image. Their outputs are collectively converted in the brain to form a visual sensation. Such a mechanism of simultaneous image collection was impossible to imitate and, after some artificial attempt without practical results, was immediately replaced by the point-by-point sequential exploration called *scanning*.

The original scanning system was implemented and radio wave transmission thereof for the first time after the *First World War* by the Scottish inventor *John Logie Baird* (Fig. 1, b. Helensburgh, August 13rd, 1888 – d. Bexhill-on-Sea, June 14th, 1946). On October 2nd, 1925, Baird in London did the first practical demonstration of a television transmission system created by himself, sending at distance a real and typical television image composed by 30 lines: **TELEVISION** (Grecian word for 'vision at a distance') **WAS BORN**! One of the images so transmitted was his office worker *William Edward Taynton* thus becoming the first man in history to appear on television.

Baird's television scanning system was a mechanical system. A *Nipkow disk* turned in front of selenium sensing elements and sequentially creating an electrical signal which followed the brightness of every image dot, line after line. The display consisted of another Nipkow disk (rotating in synchronism with the distant first) that turned in front of a neon lamp now controlled by the first signal and modulated according to the brightness of the original dots. Some years later, Baird also experimented with a rudimentary mechanical and not very practical colour transmission system but it was useful to gain experience.

Successive images, slightly different from each other, and importantly within a controlled limit of exposure rate, would be overlapped on the retina, giving the sensation of movement: this is the mechanism for projecting a film in which the shutter is periodically interposed between light source and the film when it is switching to the next frame.

Moreover, here there was a first huge order of difficulty that such a system would involve. Assuming approximately the same number of frames per second in both cases, in the same interval in which the light beam of the cinematographic machine projects a frame onto the screen, the scanning system must complete the exploration for points of the relative image and the receiving medium its reconstruction.

The Baird's system was inadequate for performing a clear and clean vision. In those experiments, the rate was around 5 fps (frames per second) in opposite to 24 fps of the cinematograph at that time in vogue and so some past opinions said that the television would have had a short life. Further development of Baird's mechanical scanning system did not bring sufficient further improvements and it was abandoned in favour of the electronic scanning system, i.e. the image is scanned by an electronic beam of a special image-sensitive electronic tube invented a couple of year before and dramatically improved meanwhile. This beam is an intangible agent; it is made up exclusively by negative electric charges (electrons) and because of its very nature is devoid of mechanical inertia. The electronic beam can be easily and instantly deflected using electrostatic or magnetic fields according to the two movements prescribed by the needs of image exploration: horizontal shifting during the exploration of the points of a row and vertical shifting for exploration of the lines and the subsequent return to the starting position for processing the next image.

The image is thus ideally divided sequentially in a series of thin horizontal stripes (lines) immediately close to each other. So each row is analyzed according to the sequence of elements (dots or *pixels*) to which the mechanism of exploration equates an ordered series of electric pulses directly proportional in intensity at the grade of light of each individual element. Therefore, the impossibility of a synthesis, during the transmission, is substituted by the possibilities of analysis; the contemporaneity of the image is artificially replaced by the sequence of those elements that compose it in the very short interval in which it can approximately be considered static.

When received, the image is then reconstructed on a display (an electronic cathode tube invented and improved in the meanwhile too) in the succession of individual items according to a sequence identical to the transmitted ones, using special timing pulses 'merged' in the transmission, giving a contemporary visual sensation, thanks to the well-known phenomenon of image persistence on the retina. So luminous pulses alternated with darkness make a continuous visual sensation as long as they occur with a frequency greater than a certain value, being higher the greater the brightness of the image.

This methodology of the scanning system for sequential points has remained virtually unchanged until today.

Television, through the electrical signal intermediary, is thus technically not a pure optical phenomenon in the normal meaning of the term because the image light has been converted

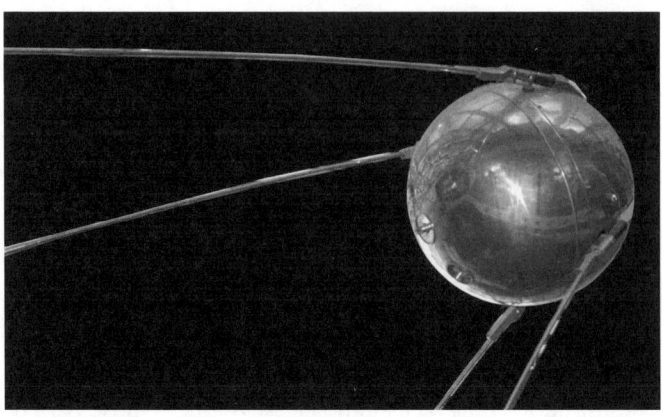

Fig. 2 - A replica of Sputnik 1, the first artificial satellite in the world to be put into outer space

into electrical signals that, via radio waves, travelled distances to be reconverted as light at the receiving place. However, due to the high speed of propagation of radio waves (300,000 km/sec), this can be virtually at the same instant as the original image was captured.

Thanks to the progress of technology, the spectator seated in his house sees, mostly even in real time, what happens in the world exactly as it is happening. And not only in the world as we know it! Let's think about the exciting moments of the first steps of a man on the Moon. Or the spectacular pirouettes of the astronauts inside the space shuttle or in the vacuum of space. Or the images sent daily from Earth to the many satellites that orbit around (Fig. 2). Those images are broadcast on various parts of the globe instantly!

Thanks to television the eye has exceeded the narrow human limits of man's ability to see. It can now explore safely anywhere, reaching the farthest recesses of the world that otherwise we could never dream of seeing.

Baird likely did not fully comprehend that television would become one of the most revolutionary phenomenon of history, able to radically transform conceptions and habits of millions of people!

The civilization of writing and speech was therefore replaced by the civilization of the image and the television is the emblem. Of course, this revolution brings some negative aspects, because if the television on the one hand has literally brought 'the world at home', on the other hand it has transformed the spectator as a passive object, easily dependent on the message launched by the television medium.

However, thanks to the prodigious progress made by technology, the television has in recent years provided a formidable sample of its exceptional ability having transformed itself, from a simple intermediary between reality and the spectator, in medium of creative interpretation of reality from a passive object. In short, the spectator has found the possibility to become an active player for making his own television.

How?

It was electronics that, through an always constant evolution, has made possible the design and the construction of the totality of the television equipment: recorders, cameras, mixer, titling systems, etc.

An increasing production on a large scale and the possibility of using a form of lease or rental, gradually made available to the general public such devices, some of them with the opportunity to self-build with relevant projects presented in several magazines specialized in electronics.

To know about this equipment, to see and understand how it works, to build it and to learn to use it for personal use correctly, may open new horizons to our cultural preparation and our opportunities to enhance our spare time in intelligent and fun ways.

Everything turns around this fascinating word: **VIDEO**.

1. The Universe of Video

VIDEO is a Latin word meaning '*I see*'. It takes first place when speaking about the electrical signal that contains, among other things, all the informations about an image, the so-called *Video Signal*.

As previously mentioned, in a telecast the camera systems generate the video signal sent to the transmitter that will transform it into modulated electromagnetic waves. The receiver, in order to convert such waves into video signals, allows us to view, almost simultaneously, what they are framing. However most of the current programs beamed by any TV station are not subject to this principle of simultaneity, because they are not broadcasted 'live', but were previously recorded to be broadcasted later.

Precisely it is the set of 'recorded images', its production rules, its language, the equipment used and so on, which is generically referred to as **VIDEO**; not only the recording or playback of images with synchronized sound, but all the information about generation, production, processing, assembling, testing, recording, playback and viewing television images.

Until a few years ago, 'working on video' was exclusively reserved for the professional experts because the equipments' costs were very high and their structure quite complex. Today, thanks to the research and discoveries made by electronics, video is within everyone's reach, with high quality equipment, simple to use, applicable to different uses, and above all at reasonable costs.

The use of video does not end with the recording of television programs or with the vision of the films offered by rent & sale organizations of pre-recorded videotapes. The recording of images and sounds on a magnetic tape makes video systems an irreplaceable element both in the field of entertainment as those in communication, teaching, research or medicine.

From its beginning the amateur video system was compared with 'home-made' cinema movie and so the 'fight' of the magnetic tape against the film was inevitable: the result was that portable video recorders during the years have rapidly replaced the *Super8* movie cameras (Fig. 1).

Fig. 1 - Braun Nizo 800 Super-8 film camera (Courtesy *Hannes Grobe*)

The immediate control of recorded images, the ability to erase and reuse the same tape many times and its longer duration compared to Super8 film reel were subjects more than sufficient to bend the needle of the balance in favour of the video system. A number of people immediately recognized the great potential for social changes inherent in the use of video portable equipments. People such as the exhibitionist for whom the video system was 'a dream come true' or the social documentarian who could record many hours of conversations with the man in the street, for personal interest or work.

Lots of people were quickly consumed by the 'video bug' and this was only the beginning (Fig.2)! Moreover, we must consider that films suffered a heavier deterioration over time compared to magnetic tape. Also, most film industries had the opportunity to store film in climate controlled (temperature, humidity, lighting, etc.) rooms, but the average user did not often have this prerogative.

Fig. 2 - Man recording images of his wife and daughter

The duplication of film did not solve the problem because the process of copying, even if it returns a new support, would introduce a further deterioration of vision due to the copy itself.

With digital electronics even the tape was gradually replaced with digital media such as optical discs which have generally a long lifetime and it is always possible to make identical copies of the original.

However, films and tapes have the advantage of a complete and well-established standardization, which make them very suitable as visual assets to be stored for long times vs. the great variety and rapid evolution (and the consequent obsolescence) for most video digital recording formats and supports.

Just a few years ago, most amateur & professional video equipments were designed primarily for analogue video. The applications for video were somewhat confined into analogue broadcast and cable television, analogue VCRs, analogue set top boxes with limited functionality, and simple analogue video capture cards for personal computers.

Until recently digital video was confined to professionals, such as video editors. But the relentless increase of microprocessors' computing power and the capacity of memories, with continuing falling costs of personal computers over time, means that digital video technology has become the first choice for the average consumer.

This trend has led to the development of specific dedicated digital equipment so there has been a tremendous and rapid conversion to digital video, mostly based on the MPEG digital video standards and the ability to use the internet for transferring video data.

But in the specific context of this book, we will discuss almost exclusively about *Analogue Video Equipment*, introducing the electronic machinery responsible for the generation and processing of the analogue video signal. We will examine specific pieces of analogue video equipment, how they work, their inner mechanics, weaknesses and danger signals. Beginning at the very basic levels, we will work toward a greater degree of complexity.

So why do we prefer speaking about analogue video rather than digital?

The reason is because the analogue video is simpler to elaborate than digital video, which implies some specific knowledge which a typical electronic technician may not possess. Analogue video equipment is still generally considered a little more reliable due to the fact that, for example, a recording videotape system is not likely to fault like a computer hard drive, a solid state memory card or a bugged operating system, risking some or all data to irretrievable loss. This book could also represent a guide for those who need or desire to learn about video. Or anyone who wants to evaluate or simply know more about analogue video systems and has some electronic DIY skills. The projects presented inside provide all the instructions (including schematics and even some circuit board layouts) and explain clearly all the steps to obtain the right final result. A basic understanding of the technical processes hidden behind all those knobs, measurements and calibrations is essential if a specific piece of video hardware is to have meaning.

Furthermore the world as we see it is analogue. So many shades of green, red and blue combinations which excite our retina, creating moving images with a continuous and unlimited flow of visual information, are received through analogue video with a congenital delay to our

Chapter 1. The Universe of Video

sight, a difference of vision that models figures in filigree and animates them delayed in time.

On the other hand digital video can appear harsh, overly sharp, 'chequered' and 'metallic' to our sight. A number of optical and electronic or software (plug-in) filters has been expressly designed to counteract these problems.

We can make a paragon of analogue video with the music played from a vinyl record via an electronic tube power amplifier. The naturalness of the diffused music, the timbre of the instruments and the harmonic richness combined with the great dynamics of a vinyl (well cared for to give the best of itself) is incomparable versus a digital audio disk played on a solid state power amplifier.

VIDEO was one of the key elements which belong to the convergence of all the various technologies previously unrelated in home and business sectors, like multimedia, cinema, computer science, word processing, mobile, meteorology, and so on.

Fig. 3 - A nice girl who handles an amateur small digital camera

VIDEO is an electronic mirror that could allow an individual or a company to become more aware of them self and come to terms with their inner being.

VIDEO has changed dramatically from heavy and bulky video cameras & recorders in days of yore to physically light cameras which are today capable of burning a DVD or saving onto a flash memory while shooting.

And how can we forget the tiny size of a video mobile phone? It is perhaps the first to undergo a change in technology and also be accepted (or declined). The day is not far when the phone could be a test bench for the rest of the media and future of television (Fig. 3).

A last jump into the past, explaining how everything was born and evolved before the digital age, is the goal of this book too.

The *Analogue Video Universe* has represented an exciting creative adventure opening our eyes to new ways of perceiving and communicating with others.

2. Television standards

The current standard television system has been arrived at through tortuous routes and numerous stages. The high cost of trials for the introduction of new systems using ever more innovative new ideas created a considerable inertia to practical implementation. So technological changes, especially in recent times, had almost always been undertaken gradually, while maintaining a degree of compatibility with legacy systems thus introducing gradual improvements and additional services.

Let's go to see summarily and chronologically the stages that led to the standardization of the current television technology.

2.1 History and Prehistory of Television

Surprisingly, if you ask someone who invented television, they most likely will not respond.

We know that the *Wright brothers* invented the airplane and *Antonio Meucci*[1] the telephone, *Guglielmo Marconi* invented the radiotelegraph and *Thomas Alva Edison* the electric light bulb, *Alessandro Volta* the electric battery and *Louis Augustin Le Prince*[2] the motion picture camera, but who invented television?

The television was not invented after a stroke of genius as a result of experimentation. For decades it was an elusive and troubled dream of the most brilliant scientists who for many years tried to build an efficient and working prototype, and many more to improve it. From 1872 to the boom in the sixties, television was a race full of small triumphs, devastating failures and insane battles for patents and colour television transmission standards. Even the same technology has been out of control for decades! The events of the world, such as the two world wars, have delayed developments at the crucial moments.

In 1831, *Joseph Henry* and *Michael Faraday*, who were working on electromagnetism, contributed to the age of electronic communication.

In 1862, *Abbe Giovanni Caselli* invented a machine called a *Pantelegraph* which could be called the ancestor of the modern fax machine, and became the First person capable of transmitting an image through cables at distance.

In 1869, *Johann Wilhelm Hittorf* performed experiments with selenium and light. He discovered that, by modulating the light that struck the selenium, it might modulate an electric current running through it. In other words, when the light increased, also the current increased and when the light dimmed, the signal decreased. This discovery opened the way to inventors for the transformation of images into electronic signals.

A few years later, *Ernst Werner von Siemens* set up the first selenium photocell and in 1876 *George Carey* began to think about a complete television system. The following year he showed the drawings of what he called a 'selenium camera' or *Telectroscope*, which would have allowed people 'to watch' through electricity.

Fig. 1 - Paul Julius Gottlieb Nipkow

Chapter 2. Television standards

Eugen Goldstein coined the term *Cathode Ray* to describe the light emitted when electric current is 'pushed' inside a tube.

In 1884 *Paul Julius Gottlieb Nipkow* (Fig. 1) was able to send pictures through wires using a rotating metal disc. This new technology was called *Electronic Telescope*, better known as the *Nipkow disk*, which the first television electromechanical prototypes were based on.

Fig. 2

By arranging some holes in a progressively outward pattern on an opaque disk, and running the disk, you were able to analyze the images row after row, starting from the outer hole to the inner. From now on, we heard about 'television lines'. The Nipkow disk had 18 lines of resolution.

The receiving screen, half the size of a common business card, was composed of a small motor with a disc and a neon lamp which worked together releasing a red-orange light.

At the *World Fair* held in Paris in 1900, the first *Congress of Electricity* also took place, where the Russian *Constantin Perskyi* used for the first time ever the word *"Television"*.

After the First World War, *John Logie Baird* was able to reunite the ideas of his predecessors. The eccentric Scottish man created the first prototype of mechanical television in 1925 by sending a television picture consisting of 30 lines (Fig. 2) over a distance.

His previous inventions (including a thermal undersock called *Baird Undersock* and a rust-resistant glass razor) had limited success, but he enjoyed fame and fortune with his *Televisor*.

In the same year the Russian *Vladimir Zworykin* had another idea: he put some light-sensitive elements on a small sheet of mica (an insulating compound) and coated the other side with a layer of silver, forming a kind of capacitor, enclosed inside a vacuum ampoule. On the front of the sheet he focused an image through a lens in a similar manner to that already widely used in photographic cameras. A beam of electrons could read the image line after line obtaining in this way a 'video' signal: he had invented the **ICONOSCOPE**.

For a long time the two methods (electrical and mechanical) coexisted. Baird's television, being based on a mechanical scanning system, was unaware of the new technologies. Baird saw his dream vanish so slowly at the expense of the wholly electronic system which gave an image with more lines and ultimately prevailed.

Philo Taylor Farnsworth was the first person who made the world's first working television system. It had electronic scanning of both a pickup device designed by himself called an *Image Dissector* and a display device, the first electronic television set of history (1927), surpassing the previous electromechanical one realized only two years prior from Baird. The Farnsworth's TV set projected the images on a sensitive surface using a cathode ray tube based upon *William Crookes*'s studies about *Cathode Ray Tubes* (*CRT*) and put in practise by *Karl Ferdinand Braun*. The basic operation was very similar to what we can still find on our cathode ray tube television sets.

Farnsworth was only twenty years old when he demonstrated his invention to news media on September 1st, 1928, televising a motion picture film and to the public at the *Franklin Institute* in Philadelphia on August 25th, 1934, televising live images. He had already started working on this six years before!

Nevertheless he, perhaps the youngest and most ambitious of the 'lone' inventors, had to be beaten by the scientist Vladimir Zworykin who paved the way for the mighty RCA (*R*adio *C*orporation of *A*merica).

Zworykin's electronic system lacked some of the subtleties of Farnsworth's (on which he was inspired causing some patent infringements), but his genius, combined with the financial support of RCA, was full of relentless successes.

In the same period *Charles Francis Jenkins* did some rather undocumented experiments with electromechanical television and in 1928, the *Jenkins Television Corporation* opened the first television broadcasting station in the USA, named *W3XK*, but that soon went bankrupt being based on mechanical television while the current trend was towards the fully electronic television.

Fig. 3 - A huge camera employed in Olympic Games in Berlin (1936)

In the USA in 1929 the CBS (*C*olumbia *B*roadcasting *S*ystem) was founded, while in 1930 RCA in New York the First fully electronic television station was opened. RCA three years later began to use the combination of iconoscope and kinescope designed by Zworykin. It had 240 scan lines.

Also in this period there was another crucial event heralding the next evolution of television and video. This was the entrance of industry into this field of research, which led to a rationalization of efforts, an increase of resources available to researchers and a more rapid translation into commercial results. From this period onwards, in the history of television research there would be no more great individual discoveries; the names of individual scientists would be absorbed into the brands of great companies.

Some countries began to broadcast programs, on an experimental basis, with a small number of television sets available to the public (about 200 thousands sold worldwide).

One of the first regular television broadcasts was made in 1936 during the opening of the *Olympic Games* in Berlin (Fig. 3). One of the camera operators was no less than *Walther Bruch*, the inventor of the PAL colour system! During that year there was also the first commercial programs broadcast by the BBC in Great Britain. In the USA, the first programs and regular broadcasts began in 1939, during the *Universal Exhibition of New York*.

Until now all the transmissions were made using *B*lack & *W*hite (B&W) television systems of accepted quality. But, the next goal was the transition from the relative cathode haze of B&W broadcast to a joyful and much more realistic colour.

In the United States the first tests with colour signals began in 1941. The image quality was very low, but these tests helped the development of colour systems that brought fine-tuning to the ultimate system.

Many years previously, in 1928, the Scottish engineer John Baird built a prototype of colour television using a set of three filters, red, green and blue, mechanically revolving in front of a black

Chapter 2. Television standards

and white screen. Baird designed an ingenious system to synchronize the images with the filters, which showed the basic principles of colour's formation. This remained only an experiment and was cast into oblivion for many years.

One year later *Herbert Eugene Ives* demonstrated a colour television system that was similar to that first demonstrated by Baird. In 1930, he demonstrated a two-way television (or *Picture Phone*), using video telephone booths connecting the *AT&T* and *Bell Labs* headquarters buildings in New York.

Another curious inventor of colour television was the Mexican *Guillermo Gonzalez Camarena* who in 1934 proposed, when he was 22 years old, a compromise solution based on a rotating 'tri-chromatic' wheel called *Chromoscopic Adapter* designed to be easily adaptable to existing black and white television equipments.

The advent of World War II interrupted the development of television. At that time radio was the most popular media. The post-war period is considered responsible for the birth of mass television. In the USA the TV boom occurred between 1948 and 1949, while in Europe, weakened by war, the boom spread some years later.

After World War II in the USA two colour television systems appeared both technically very advanced, each one presented by the historical rivals CBS and RCA.

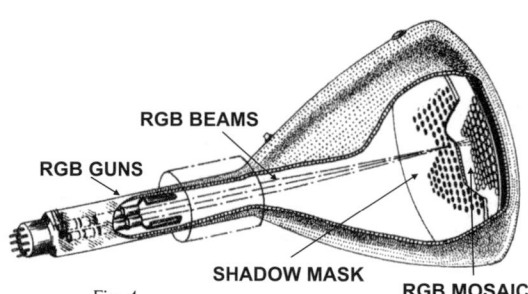
Fig. 4

But the transition from B&W to colour had to be done with minimum damage, i.e. with a system compatible with the old one, so also permitting legacy reception in black and white from a colour transmission.

All the procedures proposed for colour television were based on tri-chromatic selection; three coloured optical filters (red, green and blue) analyzed the tints for transmission and in receiving three coloured elements restored the colour image by synthesis.

The CBS mechanical system was the most perfect of its kind, but was not compatible with existing television sets. Its supporters did, however, exert pressure at the *FCC* (*F*ederal *C*ommunications *C*ommission) for its approval by saying that the RCA solution would never get past the laboratory stage. All of the RCA systems were purely electronic and one of them carried out the simultaneous transmission of three monochrome images, produced by the three colour filters and sent on three different frequencies of transmission while the receiver was composed of three kinescopes each tuned on separate frequencies corresponding to the three colours and showing the three images 'reassembled' and projected on a cinema screen. Obviously, this system was too expensive and was abandoned, but it did possess a partial compatibility with B&W systems.

Based upon French telecommunications engineer *Georges Valensi*'s patent (1938), another method employed by RCA was to implement Valensi's idea for transmitting a video signal composed of separate luminance and chrominance, so literally beating CBS engineers because it was now fully compatible to all electronic B&W systems that actually worked. The greyscale information (the so-called *Luminance*) was encoded in the usual way while the three colours (the so-called *Chrominance*) were encoded on a phase & amplitude modulated subcarrier.

Zenith Radio Corporation was given credit for contributions to the development of a special electronic picture tube, the *Tri-Colour Shadow Mask Kinescope*, which made all this possible (Fig. 4). On old B&W TV sets colour transmissions were reproduced in B&W whilst on the new colour TV sets the B&W transmission were reproduced in greyscale and coloured transmissions in full glorious tri-chromatic colour. An important step had been achieved in migrating definitely to colour transmission technology because this 'compatible colour standard' retained full backward compatibility with all existing black & white television sets.

RCA's first experiments (1950), however, were assessed for the time as unacceptable by the FCC, yet they were promising. It is in this context that the whole television industry developed appropriate rules compatible with the emerging colour television, under the auspices of the *National Television System Committee (NTSC)*. The participants gave substantial conceptual contributions to a colour system which would ultimately be acceptable and successful.

The early RCA colour system consisted of time-multiplexed sharp sampling of low-band-passed Red, Green and Blue components and the transmission of high frequency summated Red, Green and Blue information. Improvements in the three following years brought a final and stable colour system.

NTSC forwarded this standard to the FCC in a petition on July 1953 and on December 1953 the FCC approved this transmission standard, with broadcasting approved to begin next January 1954.

The NTSC television system was adopted by most of the America's countries, Japan, South Korea, Taiwan, the Philippines, Burma, and some Pacific's islands and territories. The high level of performance of the NTSC standard for almost 50 years is tribute to the individuals and corporations who developed its various concepts.

Around 1950, on the other side of the Atlantic Ocean, during the planning phase of colour broadcasting in Western Europe, it was preferred to avoid using the NTSC system due to its low compatibility with the 50 Hz frequency electricity network and also for its design problems, including poor stability of the colours during transmission problems for which NTSC was jokingly renamed 'Never Twice Same Colour'.

Two alternative systems were developed: the French *SECAM* and the German *PAL*.

A team led by *Henri de France* working at *Compagnie Française de Télévision* (later bought by *Thomson*) invented SECAM (*Système en Couleur à Mémoire*, sometimes known as *Séquentielle Couleur à Mémoire*, or even *Sequential Colour avec Mémoire*). Historically it was the first European colour television standard.

Research on SECAM began in 1956 and this technology was ready by the end of the '50s, too soon for a wide introduction, although other studies followed for improving compatibility and image quality. SECAM was inaugurated in France on October 1st, 1967.

In the opinion of eminent historical academics, the SECAM development was also prompted by a desire to protect the French television industry being confident that the NTSC system would not be popular in Europe, for whatever reasons including a lack of colour stability. It is very likely that the SECAM development was also partly justified out of reasons of national pride. Henri de France's charisma may have been a factor, coupled with the fact that France's President De Gaulle might have had some political resistance to replace a French system with one developed later in Germany, PAL.

In SECAM the colour difference signals are separately frequency modulated on two different subcarriers of frequencies, transmitted on alternate lines and stored in a steel or glass acoustic delay line in the receiver. Its advantages in robustness and immunity to phase-shift errors are outweighed by its higher visibility on monochrome receivers, poorer colour quality and greater complexity at the broadcasting end of the chain.

SECAM was far from being technically perfect yet. For example, you could and still cannot simply mix together two SECAM video signals, which is possible for two locked NTSC or PAL signals. Most SECAM TV studios use PAL equipments (in component mode) and the signal is converted to SECAM before it goes on the air. Also, the colour noise is higher in SECAM. Recording the signal to video tape is tricky, and you get poor results unless you use professional and very expensive equipment.

SECAM is still used in France and its territories, much of Eastern Europe, the Middle East and northern Africa, in some Francophone and former USSR countries, though many of the latter began to change to PAL following the fall of communism.

PAL (*Phase Alternation by Line*) was developed in 1962 by Dr. *Walter Bruch* of *Telefunken* in West Germany as an offspring of NTSC and SECAM. The format was first unveiled in 1963, with the first broadcasts beginning in the United Kingdom and Germany in 1967.

Substantially, the PAL system is an evolution of NTSC. Due to reversal of subcarrier phase on alternate lines, any phase error of a line gets corrected by an equal and opposite error on the next line because, when this signal is re-inverted in the receiver, the previous one has been previously stored in a glass optical delay line, thus correcting the original error.

A secondary advantage of adding and subtracting the direct and line-delayed signals in time is that a comb filter action is produced, resulting in excellent separation of colours.

However PAL also suffers some problems such as the loss of colour editing accuracy, due to the alternation of the phase of the colour signals which reach a common point once every 4 frames performing an editing accuracy of ±4 frames. Another problem is the line-by-line variable colour amount. Since PAL achieves accurate colour through cancelling out phase differences between two subsequent lines, the act of cancelling out errors can reduce the amplitude colour while holding the colour tint stable. Fortunately, the human eye is far less sensitive to amplitude variations rather than to tint variations, so this is very much the lesser of two evils.

PAL was adopted by the rest of the world not already using SECAM or NTSC.

Curiously in Italy, in 1972, following eight years of experimentation in a laboratory dedicated exclusively to R&D, *Armando Campioni* from *Indesit* (in collaboration with the *Seimart*) proposed a solution to the Italian Government, the television colour system *ISA* (*Identificazione a Soppressione Alternata* or *Identify Suppressing Alternately*). This was an Italian alternative to the PAL and SECAM systems already well-established in Europe. The ISA project attracted much attention and discussions but, albeit the Italian solution was technically best, it was not accepted by the Italian Government because of its non-compliance and harmonization with existing European systems. Yet it would have solved the long standing problems both for the choice of a standard and for having to pay royalties for PAL or SECAM systems. ISA was an evolution of PAL because it corrected the differential gain error on successive alternate lines through a line-by-line automatic gain add-on (US Patents #4021842 & #4134127).

In Europe, Italy was the last nation to begin regular colour PAL transmissions (1977) due overall to a short-sighted political will; Greece migrated from SECAM to PAL in 1992.

In the following years numerous sub-standards were developed but for the most part they were actually minor variations or combinations of the NTSC and PAL standards. Incredibly all SECAM variants were incompatible one with each other!

In order to convert one standard into another, some equipments called the *Standards Converters* have been designed but these are not recommended if any better solution is available. There will always be losses in the conversion and the output will not have the same quality as the source signal.

NTSC and PAL only describe the system of colour and are not the full specifications by themselves. B&W signals sometimes use these names or are referred to by their underlying standards. NTSC is based on the RS170(A) sync and timing specification. A B&W camera that is compatible with the US standard may be called NTSC compatible timing or RS170, with the latter being the more common choice. Similarly, B&W cameras destined for the UK may be referred to as PAL compatible.

The creation of a quasi-perfect solution found for a backwards-compatible television colour system was and remains an engineering, mathematical and technological marvel.

2.2 What standard to choose?

Since it was clear that the path for creating an acceptable television system was going to be based exclusively on the electronic scanning system rather than a mechanical one, it also clearly remained necessary to give technically well-defined guidelines in order to create compatible televisions. The problem must be emphasized by examining what is the official coding of television's basic elements, i.e. the set of rules and parameters that are called a **STANDARD**.

The standard for television lies in the quantitative and qualitative definitions of the various TV base characteristics that every country makes mandatory within its territory. In order to facilitate the exchange of television programs it would be greatly desirable to have a single international standard. That unfortunately was not possible to achieve, more for lack of a common political will and unreasonable nationalistic pride than for serious technical obstacles.

The invention, mass production and subsequent widespread adoption of cathode ray tubes in domestic households signalled the birth of modern television, so finally exiting from the laboratory to begin its independent life. Yet although television became a regular service we are all familiar with it continuing to offer new challenges. Therefore, if a high degree of definition, both horizontal and vertical, is a necessary requirement for a good image quality, its fulfilment has led to considerable technical complications and churn of related video equipment.

The creation of a standard has been very important not only from a technical standpoint but also for its impact on industrial production. Therefore, once the number of horizontal lines or defining the ratio between the dimensions of the receiver screen was fixed, they could not be altered with stopping production or rendering unusable the multitude of television sets already then in operation.

One can understand therefore why the various European technical committees for years have extended their researches and discussions before finally declaring new standards.

It also needed to firmly establish the most appropriate ratio between height and width of the reconstructed image so that at the end of a line scanned by the exploration system, the return to the beginning of the next line would occur exactly at the same position on the cathode ray tube receiver too.

Chapter 2. Television standards

What were the solutions that met the best two opposite classes of requirements?

And again, what was the best solution for many other questions that find their specification in the so-called standard television?

The *CCIR* (*Comité Consultatif International pour la Radio*, or *Consultative Committee on International Radio* or *International Radio Consultative Committee*), founded in 1927, issued in the post-war period the rules named *CCIR625 Standard* also named 'European' to be distinguished from the American standard (NTSC), already well-established, which had 525 lines with 60 images per second.

CCIR625 standard defines the number of lines in which each image is analyzed, the frequency of images per second, the bandwidth of the signals, the type of interlineation chosen (sequential or interlaced), the type of modulation (positive or negative), the shape and location of signal timing and various other smaller, but no less important features.

These are the main features described in detail in paragraph 3:

* *Number of lines: 625 for frame*
* *Number of frames per second: 25*
* *Number of fields per second: 50*
* *Scanning system: Interlaced 2:1*
* *Ratio between the dimensions of the screen: 4/3 (or 1.33:1)*
* *Video bandwidth: 5.5 MHz (at -3dB)*

For this reason in England, a pioneer of television in the world at that time, it did not consider appropriate, in the early days, to immediately accept or modify the old 405-lines standard in favour of the more advanced 625-lines. The same reason justified, as mentioned previously, the state of extreme confusion that dominated America about colour television. From a technical standpoint, the problem was solved there, but its implementation happened through only one system which was compatible with the legacy procedures of 20 million B&W TV sets already installed in America before the advent of colour.

Fig. 5 - BBC Broadcasting House, Portland Place at the head of Regent Street, London

With no opportunity for agreement on a single worldwide standard, the 625-lines European standard gathered the consensus of most European nations. England stood out from the agreement for the reasons mentioned above. France was considering an 819-lines high definition standard but finally accepted the 625-line standard thanks to a European agreement which bound the introduction of colour broadcasts only in CCIR625 format.

In The United Kingdom, the change to a 625-line system was more gradual and controversial.

In 1936 the *BBC* (Fig. 5) introduced the first production television system based on 405-lines, alongside Baird's 240-lines mechanical system, the latter abandoned after some months of experimentation. Suspended during World War II, the 405-lines system, also called *Marconi-EMI*, remained in use in the United Kingdom until 1985. In 1954, the BBC lost the monopoly of the English television market and the following year a consortium of regional companies founded the commercial network, *ITV*. Some of these companies proposed a variation of the colour system, based on the American NTSC system, but the BBC managed to persuade the British Government that the start of colour transmission should be postponed pending the introduction of a system with greater definition.

In 1964 the BBC launched a second TV channel to *BBC1* naturally called *BBC2* which broadcast using the CCIR625 system that older TV sets could not receive. The PAL colour coding was subsequently introduced in 1967. In November 1969 both BBC1 and ITV began to broadcast in PAL: since all the programming was now created in the new standard. The 405-lines broadcasts were still maintained for backward compatibility until the last time, January 2nd, 1985.

The so-called 'paleotelevision' of the '50s and '60s has left many pleasant memories, not only for the extraordinary quality of its programs but also for the pioneering spirit of experimentation, projected into the future despite the substantially poor means. In Europe, something similar happened also in the '70s and '80s, when television, by now an integral part of our lives, made the historical transition from black and white to more realistic colour, already in the United States since 1954.

2.3 The waveforms of synchronisms

After so much theory, let's see in detail what is this European standard CCIR625, first describing the features for early B&W transmission and in the next chapter all the changes for adapting it to colour systems.

Inside a video signal there are two constant-frequency timing reference pulses, one for horizontal scan and the other one for vertical scan, called *Synchronisms* (shortly *Syncs*), both with different widths and waveforms. In order to ensure that the image is correctly regenerated on display device, both timebase generators must be locked-in-sync. Since the syncs are combined together to form the so-called *Composite Synchronisms*, a proper circuitry in the receiver, called the *Sync Separator*, was designed to distinguish between the two types of synchronization pulses.

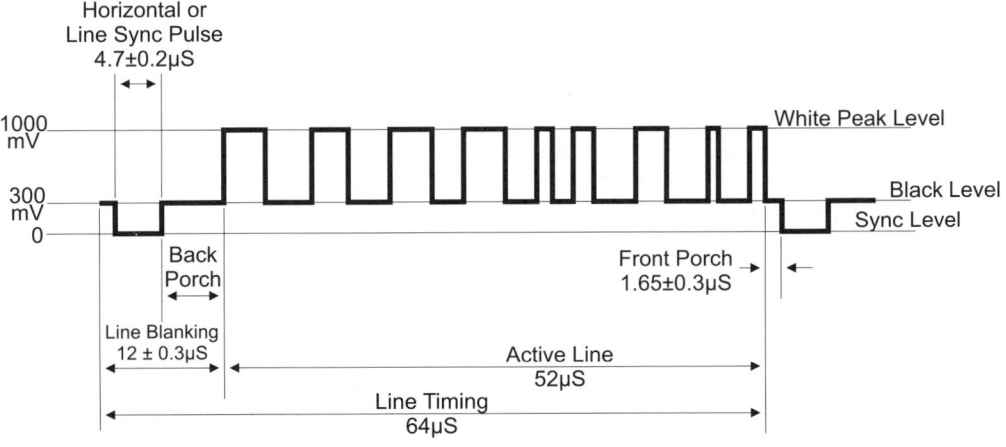

Fig. 6

2.3.1 Horizontal & Vertical pulses

In Fig. 6 is shown the waveform of one television line in CCIR625 format, as we can see it on an oscilloscope's screen. At first sight it looks a complex signal, but in reality it is not. It is composed primarily of two signals, a 'rigid' one, always placed at the beginning of each line, called *Horizontal* (or *Line*) *Sync* pulse, hereinafter *HS*, and a voltage-variable one which represents the 'active and visible' real video signal whose variations in amplitude correspond to proportional ones in brightness. It is that they produce, line after line, our image on the display device. Each HS pulse is repeated every 64µS, so the horizontal frequency is $f=1/64\mu S=15,625$ Hz.

The sync and brightness information are kept separate by assigning them well-defined signal level ranges. In particular the sync level range starts from 0V (*Ground*) to around 0.3V (*Black Level*) and the brightness above 0.3 up to 1 V pp (*White Peak Level*) for standard composite video, although the absolute DC levels may vary because usually the video signal is AC coupled. Therefore, the total range of the brightness information occupies a range of around 0.7V although it can be elevated up to 1 V in some circumstances (*White Clip Level*). This represents the maximum limit beyond which the video equipment suffers from excessive loss in performance or the signal begins to deteriorate.

Normally black level matches the same blanking level although this is not so dogmatic for other TV systems. However blanking is always present in non-picture parts of the video waveform when it is necessary for scanning system of the display device to be off so that retrace is not visible on the screen.

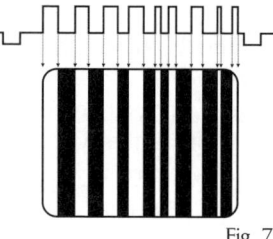

Fig. 7

The HS pulse is positioned between the active picture information and the so-called *Back & Front Porches* which improve the accuracy of sync extraction. Back & front porches timing summed to HS pulse timing usually are wide at 12µS and they compose *Horizontal* (or *Line*) *Blanking* (*HBLK*). Because each HS pulse is placed before each active line, there are 625 HS per complete frame, the same as the total number of lines.

Excluding HBLK timing from a line, it leaves the 52µS 'active line' timing. In Fig. 6 a group of square waveforms is shown which would result onto in the image on a display device shown in Fig. 7, a series of vertical black stripes on white background (or vice versa).

The Vertical Sync Pulse is quite complicated because it is composed of two types of special pulses called *Pre & Post Equalisation* and *Broad* pulses. We can see them in Fig. 8.

The five broad pulses represent the real *Vertical Sync* pulse, hereinafter *VS*, and define the beginning of a half-frame or field. To understand easily why a frame is divided into two fields we must first explain the *Interlace* concept.

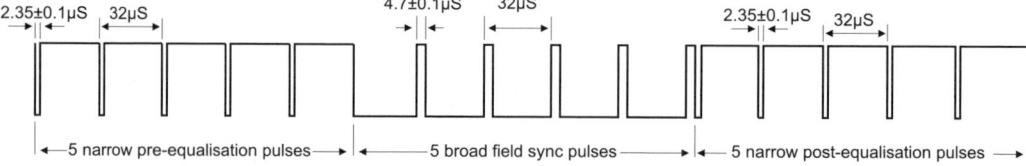

Fig. 8

2.3.2 Interlace 2:1

Interlace was a technique employed since the beginning of television and is used in all systems of analogue television broadcasting as a way to achieve good visual quality within the limitations of a narrow broadcast bandwidth, so avoiding the 'jerkiness' annoying defect.

In a cinema the film movies are shown at a rate of 24 frames per second, although 16 and 18 frames per second have been used for silent films.

In projection, during the transition from one frame to another, the projection beam is obscured. This procedure avoids the viewer seeing film scrolling and witnessing the black space between frames. However, showing pictures at this rate would produce an intrusive flicker effect which arises because our human eyes are particularly sensitive to brightness changes at a rate up to 40Hz.

In order not to perceive flicker, each frame is projected twice: first the frame is projected for a time corresponding to half the total time, i.e. for 1/48 of a second, then the beam is interrupted and then the same frame is projected for the second time for a duration of 1/48 of a second; finally the beam is again interrupted and the frame is replaced with the next. This results in 48 illuminations of image every second and therefore for the observer the brightness is stable by virtue of retinal persistence.

In television, the same effect of increasing the 25 Hz rate by a factor of 2 is achieved by the use of interlacing. Each image is divided into two fields and each of them has a single vertical scan. Each field consists of half of all the total lines of an image (i.e. 625/2), scanned at twice the image illuminations rate. The second field then completes the image by scanning the lines between those of the first field. In this way a complete picture is produced in $1/25^{th}$ of a second, but the real scan rate is twice 25 Hz, i.e. 50 Hz. So an image containing 625 lines is actually composed of two distinct vertical scans, each with 312 □ lines. This technique is called *Interlace 2:1*.

Interlacing is created very simply using a video sync generator to create an odd total number of lines in a complete frame. The field fly-back must then 'interline' the line rows on alternate fields producing the required effect. It is for this reason that all analogue television systems use an odd number of lines. Importantly, in order to obtain a perfect interlace, the waveform of vertical synchronism must be always identical.

Interlace virtually increases the image definition because at normal distances the human eye is unable to perceive rapid variations in large areas of a picture, although it tends to be rather more susceptible to picture detail such as sharp edged shapes or where there can be very large differences between the two fields.

Simple systems used 'random interlace' which means that the field and line sync generators were entirely independent, so the timings would drift with respect to each other, consequently changing the video informations from field to field in a non-predictable way.

The pattern of scanning lines covering the area of the target, or the screen of a picture display, is called a *Raster*.

2.3.3 Broad & Equalisation pulses

Broad (or *serration*) *pulses* are used to maintain the line synchronization generator running continuously and to trigger the scanning field generator. To achieve this, edges should appear every 64μS for the entire field blanking interval.

Since effectively between both field pulses which compose a full frame exists a half-line offset, it is necessary to double the frequency of the line sync during each VS resulting in identical pulse groups to be used for the field syncs. Broad pulses cover 7.5 video lines and represents the vertical sync in each field.

The other set of pulses is called *Equalisation Pulses,* inserted before and after the broad pulses, introduced for allowing the sync separator to settle and be minimally influenced by the presence of the line syncs since their width is half that of the line sync pulses, i.e. doubled in frequency to maintain someway locked the line sync generator.

The equalisation and broad pulses are always identical for both fields and are shown in detail in Fig. 9.

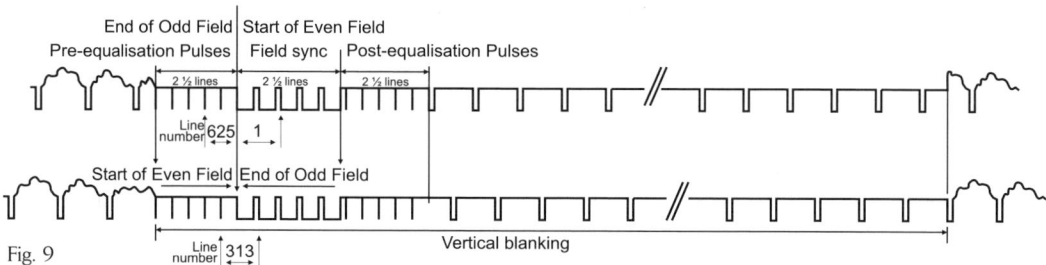

Fig. 9

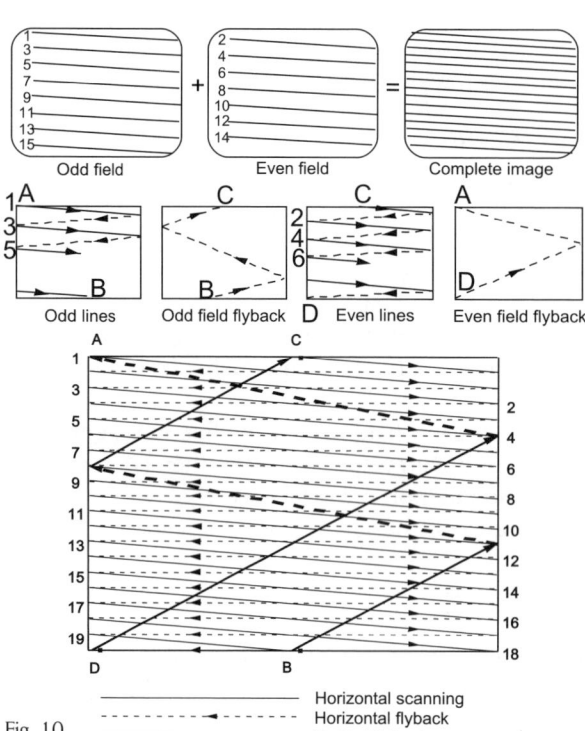

Fig. 10

Each line of a complete picture is always numbered consecutively and not as displayed in Fig. 10. However, it is necessary in order to distinguish between the two fields making up each picture, the words 'odd' and 'even' are (improperly) used. The *Odd Field* is defined as one that starts and ends in a half-line, while the other one is named *Even Field*, starting and ending in a full-line. The *Vertical Blanking* (*VBLK*) period takes 25 lines of each field, with lines generally empty (at black level), but more recently special pseudo digital information was inserted inside those lines such as teletext, closed caption or control, test and special signals.

2.3.4 Vertical frequency

The standards adopted by the American Federal Communications Commission (FCC) for monochrome

26

2.3 The waveforms of synchronisms

television in the United States specified a system of 525 lines per frame, transmitted at a frame rate of 30 Hz, with each frame composed of two interlaced fields of horizontal lines. Initially in the development of television transmission standards, the 60 Hz power line waveform was chosen as a convenient reference for vertical scan. Furthermore, in the event of coupling of power line hum into the video signal or scanning and deflection circuits, the visible effects would be stationary and less objectionable than moving hum bars or distortion of horizontal scanning geometry.

In Europe, a 50 Hz interlaced system was chosen for many of the same reasons.

In the early days of European TV, mains at 50 Hz could annoy the vision because it was sourced from imperfectly filtered power supplies. This would manifest itself on the picture by modulating its brightness or with a 'jumping' raster. If there was a difference of ±1 Hz (i.e. 49 - 51 Hz) this modulation would show up very noticeably by rolling up or down on the screen once every second. Using a 50 Hz scan caused the modulation to be stationary on the screen which is less noticeable, as previously stated for 60 Hz NTSC system.

By improvements in television receivers and in electronics, also the 50Hz power line interference was more perfectly filtered using improved power supplies, so the 50Hz (or 60 Hz) vertical frequency remained unchanged.

Fig. 11

Fig. 12

Fig. 13

All sync pulses are always at sub-black levels to be separated correctly by the video information using a special sync separator which we will discuss in paragraph ¤5, so to allow correct synchronisation of display device's deflection circuits. For this reason generally all video syncs are represented with negative polarity.

2.3.5 Aspect ratio

Once the waveform synchronisms are established, the subject of aspect ratios is considered such that the full pictures are presented on a video display device's screen.

An aspect ratio simply describes the TV screen dimensions in terms of width by height. The standard TV set is built with an aspect ratio of 4 (width) by 3 (height) or 1.33:1. This simply means that if it would be 1.33 centimetre wide if it were 1 centimetre tall. Obviously, this works for any other multiple of length as well.

The TV *4:3 Aspect Ratio* was also chosen in order to match it to a film standard defined by the *Academy of Motion Picture Arts and Sciences* after the advent of optical soundtrack on film. Before 1950, the previously photographed film could be satisfactorily viewed on TV in the early days of television medium. But, with the rapid spread of television sets, starting from the '50s a series of panoramic formats were adopted by film in order to enhance the spectacularity of images.

For this reason TV aspect ratio is important for broadcasting a movie because not all movies are filmed in the same aspect ratio.

In fact actually the most part of movies commonly use a format which owns a 16:9 aspect ratio or wider. During filming, the various frames are designed so as not to have too much important material near the edges.

During the transmission in 4:3, the edges are filled with two black bands added above and below the image (letterbox), to maintain the right proportions without making a complete conversion to 16:9 (see Fig.11, 12 & 13).

2.3.6 Video Bandwidth

The bandwidth is a parameter directly related to the maximum number of points distinguished in a video line that represents the horizontal resolution, which is defined as the number of vertical lines that can be represented throughout the screen height. This definition may seem a little rough, but particularly on television it has its justification. The higher the video bandwidth, the better is the quality of the picture.

Let's see in detail how it has come to the value of 5.5MHz.

We assume that horizontal and vertical resolution of an image should be identical. The total number of lines per image is 625 minus 2*25 (there are 25 lines lost for each field due to the Vertical Blanking) and therefore there are 575 lines actually visible for each image.

Since the 575 lines compose up a whole frame consisting of an interlaced image (i.e. two different and real half images), we must divide by 2 such lines, i.e. $575/2 = 287.5$ lines.

The television aspect ratio is 4:3, so the horizontal maximum possible number of points alternately white and blacks (building a complete waveform cycle) is $287.5*4/3 = 383.33$ cycles.

Since the visible line duration is 52 µS, into this time period therefore has to accommodate 383.33 cycles. In one second there are then $383.33/(52*10^{-6})$ cycles $= 7,372,000$ cycles, or in another way, the video information that we can show in a video line has the maximum possible frequency of about 7.372 MHz.

But this value does not match that one of 5.5 MHz used in practice! The reason has been determined by subjective tests, where some viewers, seeing real images, were asked to compare horizontal and vertical resolution by varying only the horizontal resolution.

In consequence of those experiments performed in 1933 in RCA laboratories, the engineers *Kell, Bedford* and *Trainer* determined that the effective resolution of their electronic video system was 64% (0.64) of the total scan lines when objectively they realized that beyond this there were insignificant improvements. As electronics became more capable of generating more scan lines, Kell and Bedford increased the factor to 75% in 1940. The empirical frequency of 7.372 MHz was so reduced practically to 5.5 MHz simply by multiplying it with this *Kell Factor*, i.e. $7.372*0.75 = 5.5$ MHz.

In particular, Kell reflected about the vertical resolution which obviously cannot exceed the number of active lines with the combined effects of a camera and CRT scanning trying to match the outline of the image and its vertical dimension with the TV resolution. It is evident that the greater the number of scanning lines, the better is the quality of the images. However, the ideal case where the number of active lines is also the number of vertical detail is virtually impossible to achieve because there is no way to ensure continuing alignment of the data: most of the time details will be straddled with a consequent loss of vertical resolution. The Kell factor simply denotes roughly the vertical resolution which, for the effect of its reduction factor, will be reduced to just over 400 lines, i.e. (625-50)*0.75 ≈430 lines. Because TV ratio is 1.33:1 the horizontal resolution is 430*1.33 =≈572 points. If we consider the fact that these horizontal points are distributed onto 52µS representing the timing of one active line, a sequence of white & black points can be described as the most detailed video signal possible whose period is equal to the sum of two adjacent point timings. Dividing the active line timing for half of the horizontal points, the result will be 5.5 MHz, i.e. 52µS/572/2 = 5.5 MHz which is just the video bandwidth.

Fig. 14

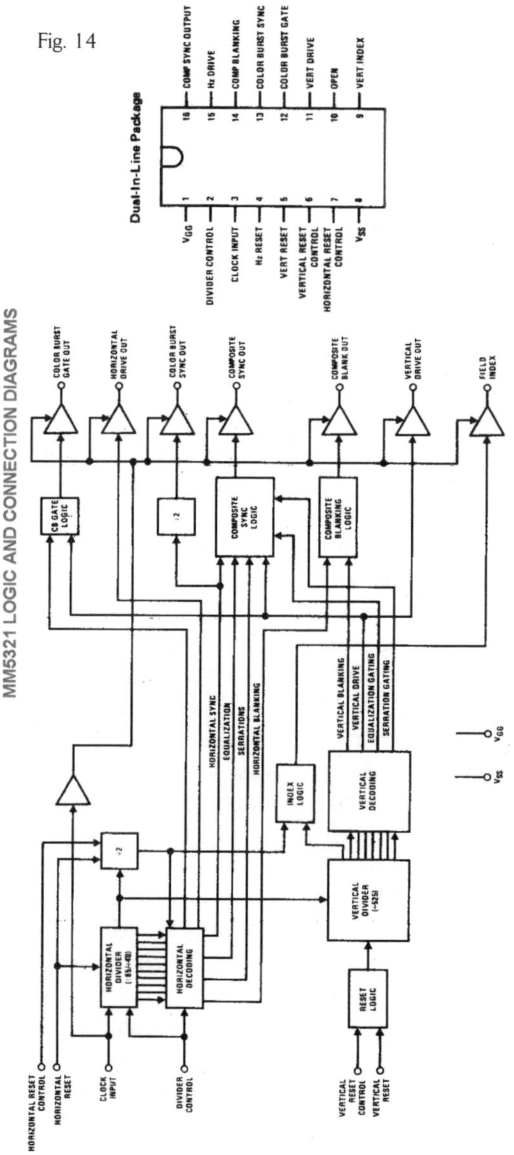

The designer of a video system knows that bandwidth is critical and has much more to consider, not just the electronics industry standard, when people are choosing a piece of video gear. On the other side a lower video bandwidth will reduce the sophistication and cost of video equipment and broadcast bandwidth. In principle all the video devices which made image manipulations have a bandwidth wider than 5.5 MHz, in order to minimize the detail loss in the video signal chain. The work of any piece of video device, after all, is to deliver to its final destination the same image that came from the original video source. Video signals are affected by bandwidth loss in many ways must be taken into account in order to make this happen.

The analysis discussed here is valid for CCIR625 monochrome systems. Other systems with different line numbers and frequency image can be treated similarly. TV colour systems have different features that need separate considerations.

2.4 Sync Pulse Generators (SPGs)

Next step is to explain how to create these synchronisms. In the early days a simple free-running oscillator built using electronic tubes with low frequency stability achieved this. As electronics evolved, other improved solutions were created, like circuits using transistorized oscillators stabilized in temperature and later TTL or CMOS ICs controlled by quartz crystal. At the end of the seventies some industries started to produce a complete sync generator in a single chip.

2.4.1 Single chip SPGs.

One of the first SPG single chips was National's *MM5321* designed only for the NTSC market. The MM5321 TV camera sync generator was a MOS, P-channel enhancement mode, LSI chip designed to supply the basic sync functions for either colour or monochrome 525 line/60 Hz interlaced camera and video recorder applications. It required power supplies of +5V and -12V. Starting from a 2.04545 MHz or 1.260 MHz input reference, the MM5321 could easily be synchronized externally, thus providing flexible control of multiple camera installations. The generator supplied the main standard sync outputs (negative polarity): Horizontal Sync, Vertical Sync, Composite Blanking (CBLK=HBLK+VBLK), Composite Sync (CS) and the Colour Burst Gate (BG) (see Fig. 14). The MM5321 is considered obsolete but some online specialized distributors in hard-to-find parts can still quite easily obtain it. However *Jameco Electronics* can still furnish it at a reasonable price (www.jameco.com).

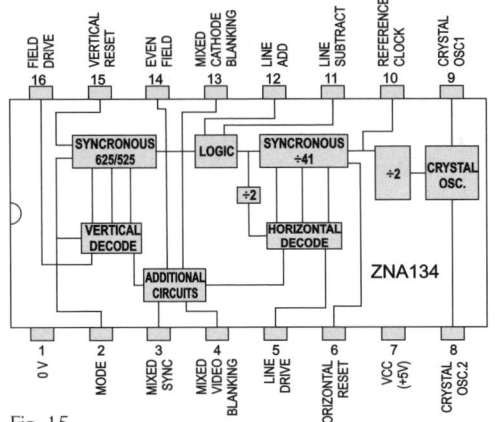

Fig. 15

Another 'glorious' SPG chip was Ferranti's *ZNA134* (Fig. 15). In its datasheet, the British manufacturer Ferranti explained the ZNA134 features thus: *"The ZNA134 integrated circuit utilizes a crystal to generate all the horizontal, vertical, mixed blanking and synchronising pulses necessary for raster generation in 625 or 525 lines commercial, industrial or military television systems. The synchronous dividers and decoding logic employed within the unit ensure perfect interlace, together with spike-free output waveforms having precisely defined relative positions and pulse widths. The device is contained in a 16 pin DIL package and can be selected to operate over the military temperature range".*

With this chip it was possible to build the simplest, full broadcast, dual standard SPG ever described. Ferranti also had been involved in production of LSI and ULA circuits used in the then home computers such as the *Sinclair ZX81 & ZX Spectrum, Acorn Electron* and *BBC Microcomputer.* The microelectronics business department was sold to Plessey in 1988 but in 1993 all Ferranti industries went bankrupt.

Apart from containing all the pulse forming SPG logic, the ZNA134 also contains an on chip oscillator designed for use with a 2.5625

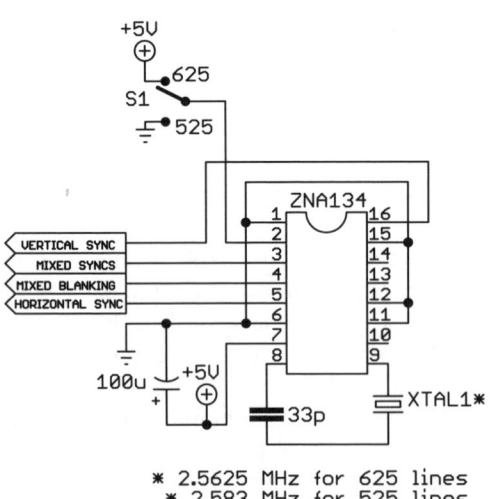

Fig. 16

2.4 Sync Pulse Generators (SPGs)

Fig. 17

(CCIR625) or 2.5830 (NTSC) MHz quartz crystal together with a tuning capacitor which for most applications can be fixed at 33pF. Also the ZNA134 can easily be synchronized for particular applications by replacing this capacitor with a varicap diode so to allow the SPG to be phase locked to other sources. Vertical and horizontal reset inputs are also provided for 'hard locking' applications.

At that time it was not a cheap IC but when you consider the number of separate ICs that it was replacing, it was possible to make a complete SPG for little money and a smaller PCB design. The ZNA134 was so 'perfect' that it was employed even in broadcast and very professional video equipments as a timing reference.

The ZNA134's application circuit is shown in Fig. 16: it is so simple that it hardly requires much extra clarification. The circuit is fed with +5V DC which must be well regulated in order not to damage the IC which 'eats' about 100mA and therefore gets quite hot: this is the normal operating condition but obviously if the supply goes over +5V DC, the ZNA134 could easily be burned.

For precise operation two slightly different crystals are required for both standards, but a crystal midway between the two crystal frequencies

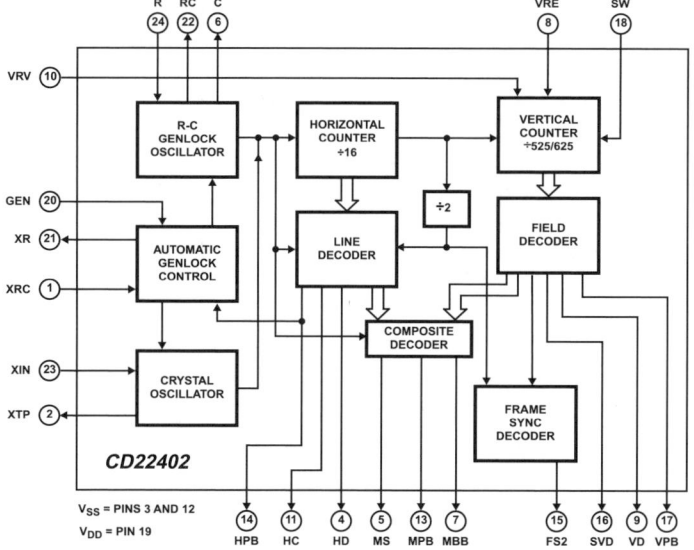

Fig. 18 — CD22402 MONOCHROME TV SYNC GENERATOR WITH AUTOMATIC GENLOCK

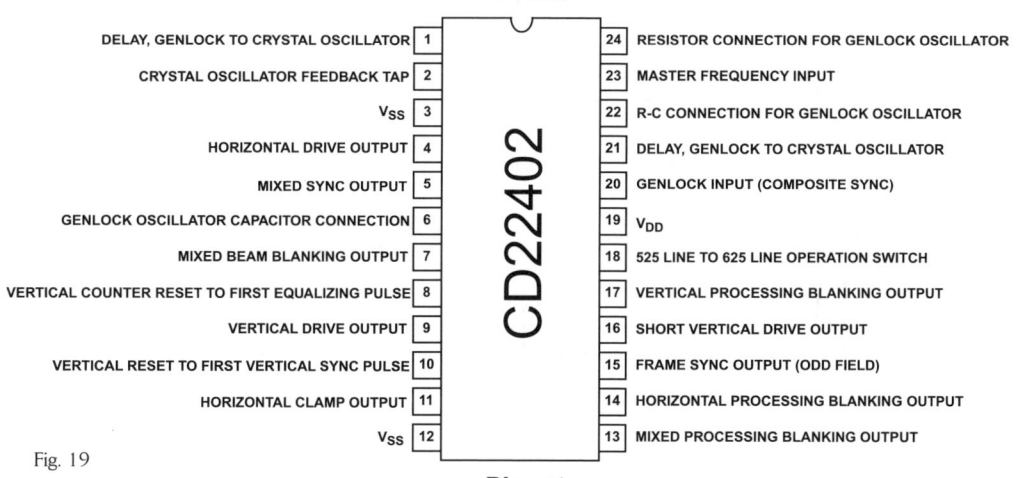

Fig. 19 — Pinout

31

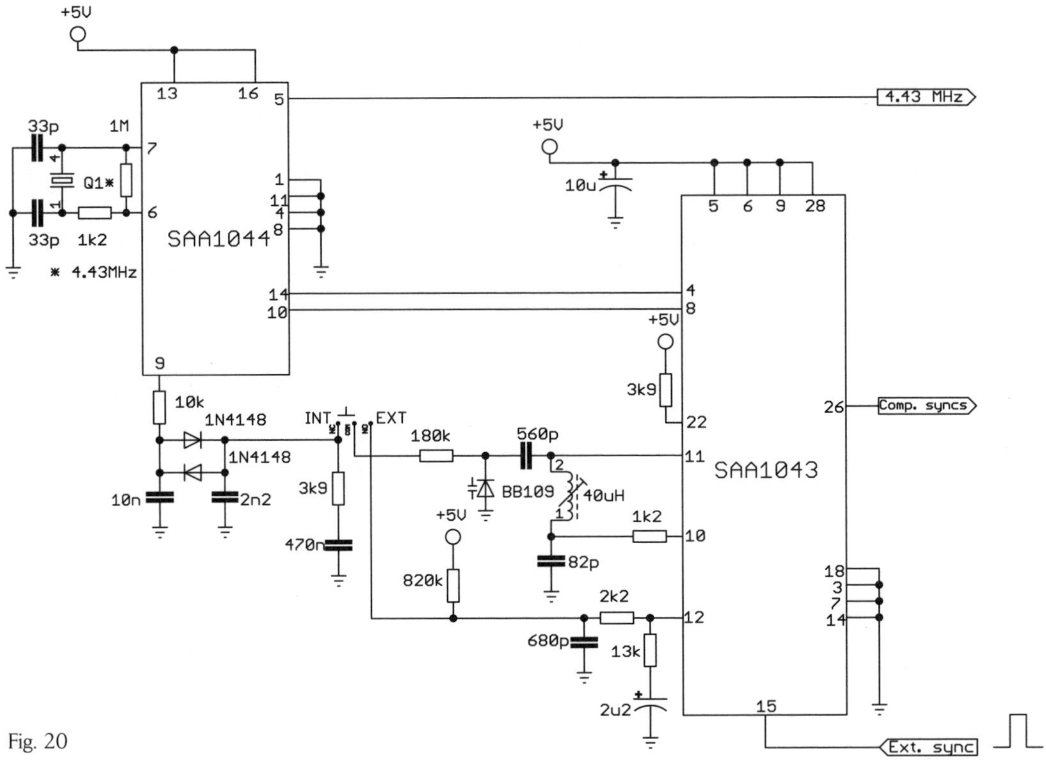

Chapter 2. Television standards

Fig. 20

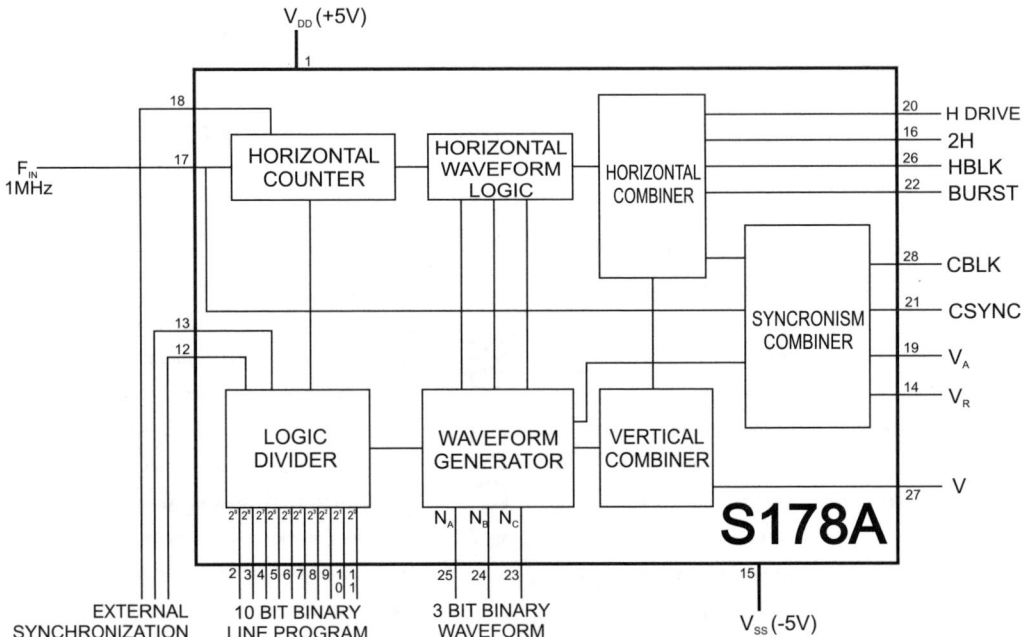

Fig. 21

2.4 Sync Pulse Generators (SPGs)

will be adequate for most applications. To change from 625 to 525 lines, a 1-way/2-positions switch is all that is required. Keep in mind that the line and field blanking widths are narrower on 525 lines compared to 625, and that the number of equalising and broad pulses is different between the two standards. All these can be seen to change as the standards switch is operated, just like the specifications say it should and they can be easily observed on a field-triggered oscilloscope screen. The construction of the unit is relatively simple so it can easily be built on a piece of *Veroboard*. In order not to damage the ZNA134 a socket is recommended for this device.

This discontinued product can easily be traced by searching on some websites of specialized distributors but I personally found it just by asking a popular auction website shop or from www.littlediode.com (Fig. 17).

The Harris *CD22402* (equivalent to *NTE7049* from NTE) is basically a 525-line SPG and it can produce a correct 525-line waveform (which has 6 equalisation and broad field sync pulse) but produces 6 of each in the 625 mode rather than the correct 5. In practice, this defect is unlikely to be a real problem but it looks like a silly oversight by the designers. It is described as a *"Sync Generator for TV Applications and Video Processing Systems"*. It is obsolete, but some may still be around. Below is presented its block diagram and pin out. Like other SPG chips, the CD22402 can be externally synchronized (Fig. 18 & 19).

Between 1988 and 1990 Philips has produced three interesting SPG chips: the *SAA1043*, *SAA1044* and *SAA1101*. These ICs are programmable to provide 625 or 525 line standards (correctly) and signals for PAL, NTSC or SECAM encoding. Interlace can be switched off as well.

Defined as a *Universal Sync Generator* (*USG*), the SAA1043 produces line drive, field drive, mixed sync, mixed blanking, burst gate and PAL switch, among other things. By selecting 3 inputs either high or low logic level, PAL, SECAM, NTSC and PAL-M signals can be generated, and also a 624 or 524-line standard for video games, omitting the half line in each field. For 625 lines a 5MHz crystal can be used, while a 5.034964MHz crystal is needed for 525 line, but it can also use an LC oscillator otherwise. The SAA1043 also includes the circuitry to lock on to an external sync source, which does not need to be fully broadcast standard compliant, as for example from a VCR source, although for this mode the LC oscillator circuit is preferred to the crystal one. The optimal power supply is about 6V and the output pulses are at CMOS levels. The SAA1044 is a SAA1043 companion called the *Subcarrier Coupler* that enables the colour subcarrier and line frequencies to be locked together. In Fig. 20 there is a typical ap-

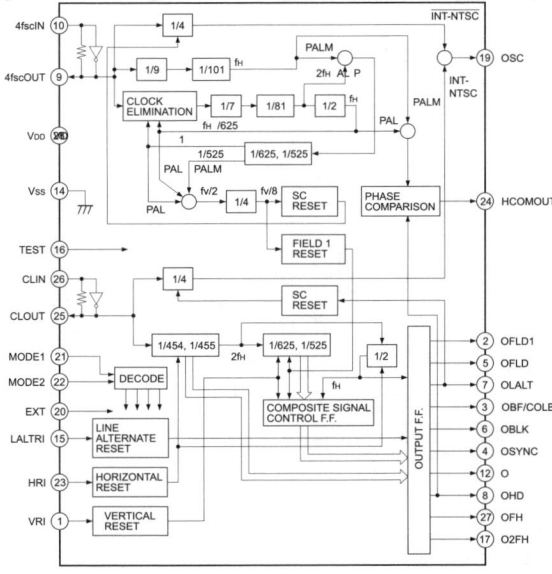

Note) Pin 19 output is (a) a signal based on Pin 26 in INT mode at NTSC.
(b) each signal is based on Pin 10 in other modes.

Fig. 22 CXD1217M

Chapter 2. Television standards

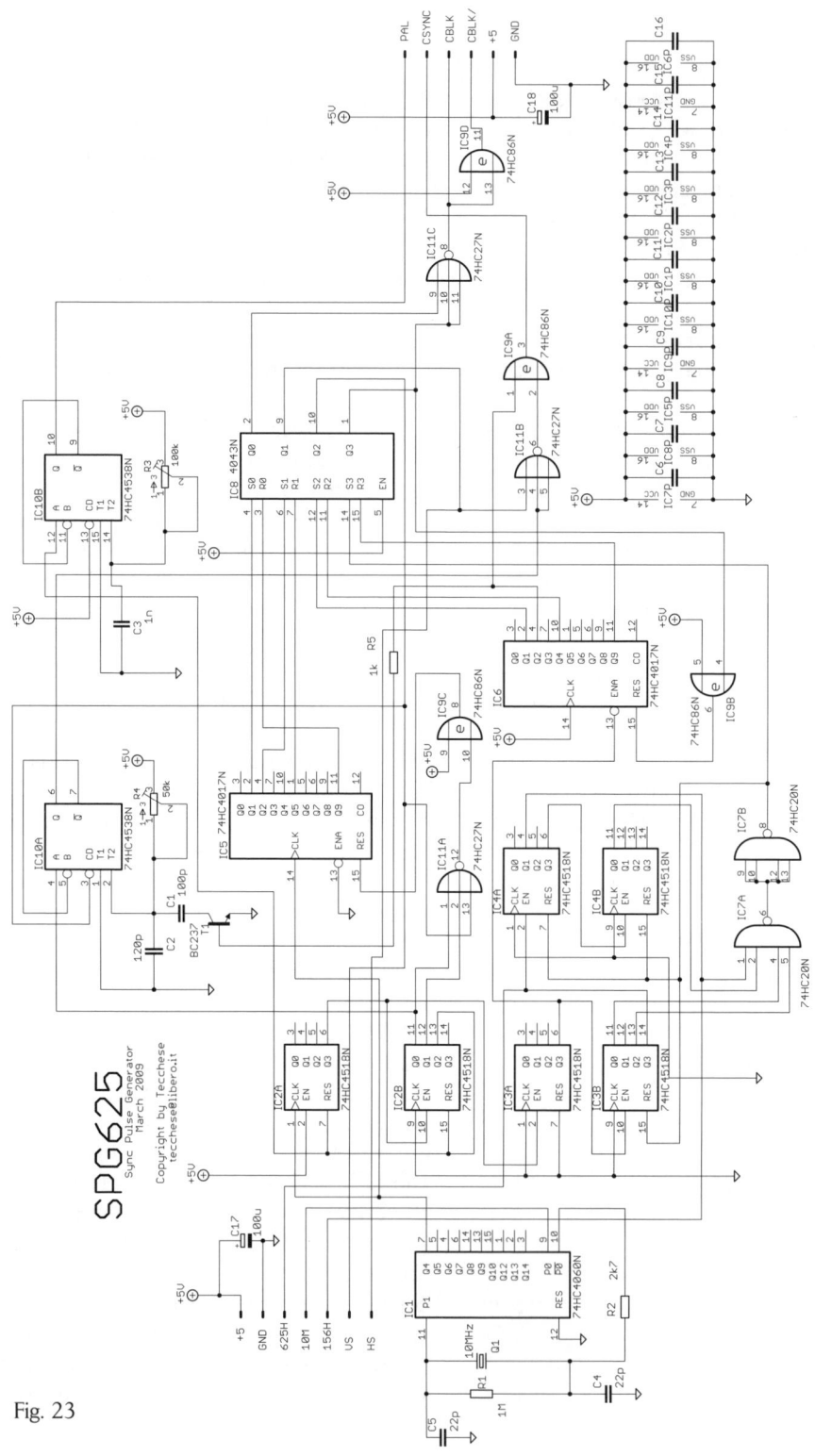

Fig. 23

plication of both ICs for PAL-ready CCIR625 system.

In January 1990 Philips introduced the SAA1101, considered as a successor to the SAA1043 sync generator and the SAA1044 subcarrier coupling IC. This device is now discontinued but it is still available from *Donberg Electronics LTD* (www.donberg.ie) or *Littlediode* (www.littlediode.com). For knowing its improved features, you can cast a glance to its datasheets still available on Philips's website.

The German Siemens also designed a SPG chip, the *S178A*, a MOS digital component which generated all the pulses necessary in a television through internal dividers, starting from a high clock frequency. You could also program all the numbers of lines up to about 1500, as well as six different sync waveforms. 10-bit binary line number codification inputs are available for odd numbers of rows (interlaced system), or for even numbers of rows by translating the numbers of those lines you want to see in its equivalent binary number. With inputs N_A, N_B and N_C, the functions dependent on standards, such as number of equalization pulses, are adjustable. It was distributed in a 28 pin DIL package and in Fig. 21 there is its block diagram. It is still distributed by Littlediode but it is quite expensive.

Even Sony made some SPG chips, most of them designed for broadcast video cameras as the *CXD1159AQ, CXD1217M, CXD1254AR, CXD1257AR, CXD1261AR*. In Fig. 22 there are the block diagram and pin configuration for CXD1217M.

Other industries made single chip SPGs as Hitachi (*HD44007A*), Panasonic (*MN67603NS* and *MN67621F*) and NPC (*SC6433* - NTSC only).

When technology made it possible to create a simple, full broadcast, multi standard SPG with a single PIC microcontroller suitably programmed (also capable of generating computer video frequencies and patterns), the production of the SPG chips was slowly stopped.

Currently in production there remains only Fairchild's *74ACT715* (available from Farnell), described as a *Programmable Video Sync Generator* and equivalent to the discontinued National's *LM1882*.

This is a wholly programmable device which can produce almost any set of video timing waveforms. It is designed to give NTSC 525 line waveforms by default but other line system waveforms can be generated. It requires some ingenuity to achieve this because registers need to be loaded with the correct numbers every time it is switched on and Fairchild does not generally reveal any application notes for performing this operation.

The 74ACT715 datasheet can be easily downloadable from Fairchild website (www.fairchildsemi.com).

2.4.2 Multiple chip SPGs.

Until now we talked only about single chip sync generators, but is there a possible means to design a CCIR625 compliant generator using only standard ICs?

The answer is affirmative and in Fig. 23 there is an example of diagram. I named this circuit **SPG625**.

This equipment should preferably be constructed by those with some experience of digital circuits and own an oscilloscope, because the circuit is quite complicated.

This unit has been built using 'only' eleven standard CMOS ICs. Thanks to their exclusive use, the power required by this SPG is low and therefore highly suitable for other video applications for which low overall consumption instruments are required.

Unlike most simple solutions based on ordinary ICs, this circuit also generates pre & post equalization pulses, broad pulses, composite blanking signal, switching PAL pulse and interlaced fields, as well as composite vertical and horizontal sync pulses starting from a unique master clock.

All waveforms are obtained by a train of pulses 1.6µs wide derived by dividing the 10 MHz crystal oscillator by 16 through the oscillator/divider IC1 (74HC4060). The 1.6µs pulses leave Q4 of IC1 and are applied to the inputs of clock of IC2A (74HC4518) and IC5 (74HC4017).

IC2 is a dual divide-by-10 counter, connected in such a way that once it reaches the number '40', both counters are cleared and counting reverts from zero.

The triple-input NOR IC11A (one of the three included in 74HC27) followed by exclusive OR IC9C (one of the four included in 74HC86), configured as an inverter, enables counter IC5 only when the IC2 is counting between 0 and 9, that is when the outputs of Q0 and Q1 of IC2B and a pulse coming from IC8/pin10 (see later about this last pulse) are at '0' logic level. Since a complete counting cycle of IC2 is 40*1.6µs=64 µs, you can obtain, according to the outputs of IC5, the horizontal signals HS and HBLK in this way: the next 1.6µS pulse sets the output Q1 of IC5 to logic state '1'; another pulse brings Q2 to the same level, and the cycle continues until IC2B reaches the number '10' and Q0 reverts to '1' and remains in this state. Then the outputs Q1 and Q9 of IC5 will drive the S0 and R0 inputs of one of the four NOR S/R Flip-Flop included in IC8 (4043) so from its output Q0 you get HBLK whose length can be calculated by measuring the timing from the two rising edge pulses coming from Q1 and Q9 of IC5, i.e. 1.6*8=12.8 µs. The 0.8 µs over the 12 µs blank timing specified in CCIR625 standard will not be critical in most cases. Similarly the Horizontal Sync is obtained from Q1 of IC8 by setting and resetting another of the four NOR S/R Flip-Flop included in IC8 through the outputs Q2 and Q5 of IC5, thus creating a pulse of 1.6*3=4.8µs. This HS pulse is delayed inside HBLK by 1.6µs, so it will be put inside horizontal blanking at the right place.

From Q3 of IC2A the pulses of width 1.6*10=16µs are then applied to IC3A (74HC4518). Since IC3A is a decimal divider, the pulses that appear on Q3 of IC3A have an interval of 160µs and are sent as clock pulses to IC6 (74HC4017) and IC3B. The decimal counters IC3B and IC4 (74HC4518) are linked together through the AND gate IC7 (74HC20): in this way, once the counter reaches the number '125', the counters IC3B, IC4A and IC4B are cleared. The counting cycle therefore will be 125x160µs=20ms: this interval matches the length of a field. To perform this operation perfectly, it is better to match IC3 & IC4 using the same brand and even production stepping if possible. The reset pulse coming from IC7B sets the S3 input of the third NOR S/R Flip-Flop included in IC8 whose output Q3 will feed IC9B (used as inverter) so that the counter IC6 is left free to work after the end of the reset signal pulse coming from IC7B.

The counter IC6, driven by a 160µS pulse train coming from Q3 of IC3A, after 8x160µs=1.28ms, reaches the state where Q9 is '1' and therefore the NOR S/R Flip-Flop will be reset via R3 input. Only a new reset of the counters IC3B, IC4A and IC4B will switch the flip-flop again. The signal from 1.28ms, and taken from the S/R flip-flop, will then be used as Vertical Blank which corresponds to 20 lines. The missing 5 lines needed for matching the CCIR625 standard are irrelevant for our final result.

The Q2 output of the last NOR S/R Flip-Flop included in IC8, through the outputs Q1 and Q4 of IC6, provides a 3x160µs=480µs pulse (corresponding to a vertical interval timing of 7.5 lines) which gates the monostable IC10A (74HC4538) and will reset IC5 through IC11A & IC9C.

2.4 Sync Pulse Generators (SPGs)

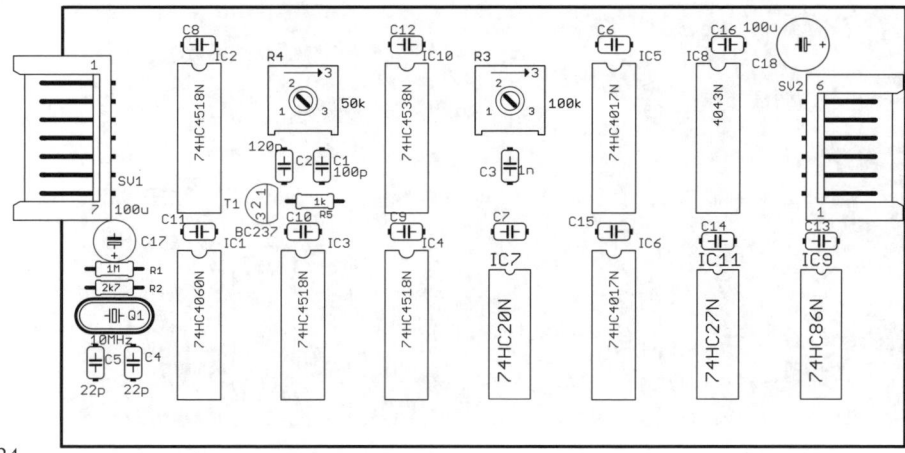

Fig. 24

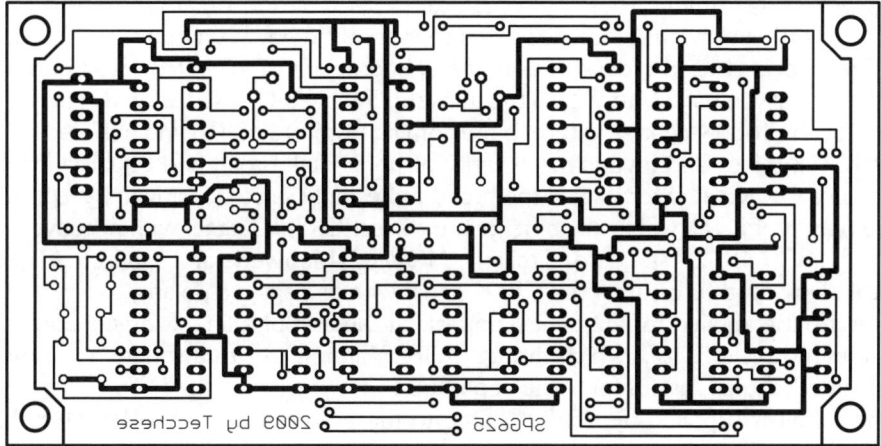

Fig. 25

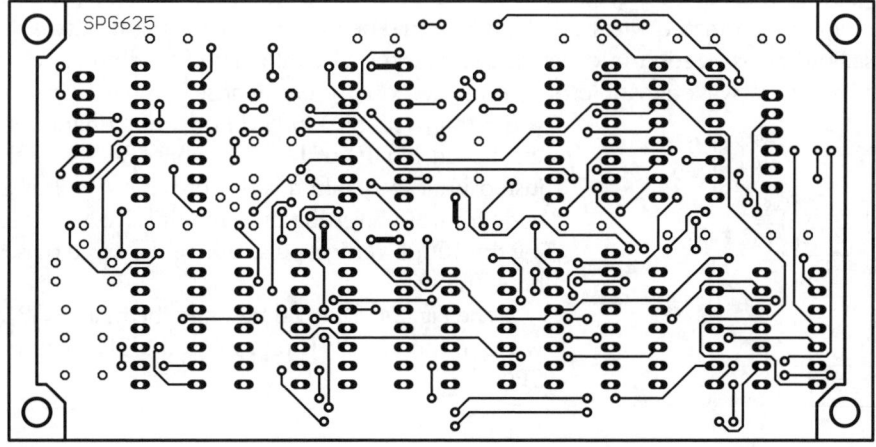

Fig. 26

Chapter 2. Television standards

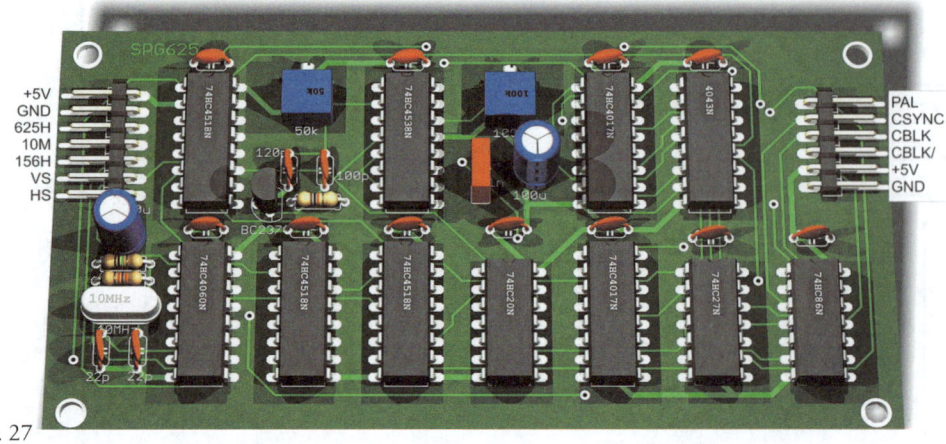

Fig. 27

The monostable IC10A is fed by pulses from Q0 of IC2B producing a pulse of 2.35µs every 32µs via R4/C2 timing. These pulses appear in place of horizontal sync during the vertical interval of 480µs previously described and represent the equalizing pulses. To enlarge the pulses inside vertical sync, in order to create the right duration of 4.7µs for broad pulses, C1 is inserted dynamically in parallel to C2 via the transistor T1 (BC547) used as an electronic switch driven only by the 160µS 'central' vertical pulse coming from Q2 of IC6 and buffered by IC9D.

NOR IC11B mixes and inverts the polarity of horizontal, broad and equalization pulses, while exclusive OR IC9A inverts again the broad pulses only since it is driven by Q2 of IC6.

At the output of IC9A will finally emerge the correct CSYNC signal with negative polarity, compatible with CCIR625 standard. Similarly, NOR IC11A inverts and mixes the horizontal and vertical blanking, thus creating the CBLK signal with its negative polarity.

The reset pulse applied to IC2, which appears at the end of a line, is drawn and used for driving the monostable IC10B, which through R3/C3 timing creates a square wave two lines wide (PAL switch signal), useful for some applications.

All the circuit is supplied by +5VDC. Due to the presence of several HCMOS IC, it is preferable not to exceed that voltage.

I designed a small single sided PCB; the top layer has a few small bridges which can be realized using some insulated wires. All the ICs should be put on sockets if simplicity is the rule.

An oscilloscope is needed in order to calibrate the SPG625. The R3 trimmer will be adjusted until a square wave is visible on the oscilloscope screen whose positive cycle must be wider than 1µS of standard horizontal timing, i.e. 65µS (probe on pin 10 of IC10B), whilst R4 trimmer is adjusted until to obtain a series of narrow pulses 2.35µS wide for 480µS (probe on pin 6 of IC10A), except during the 160µS vertical pulse where the same pulses must be 4.7µS wide.

The Fig. 24, 25 & 26 describe the main PCB layouts while in the Fig. 27 is represented a 3D picture of the SPG625.

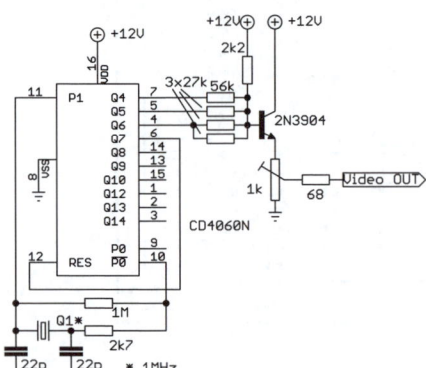

Fig. 28

2.4 Sync Pulse Generators (SPGs)

2.4.3 Non-standard SPGs

It is possible to also build SPGs slightly out of CCIR625 standard. They are used mainly for TV set calibration where a full broadcast pulse generator is not required because modern TV sets own particular sync separators able to lock imperfect video signals as might be generated for example by a video recorder or a troubled air broadcast.

In Fig. 28 there is the simplest non-standard greyscale pattern generator. It employs an oscillator/counter IC (4060), 1 MHz crystal quartz, a buffer transistor and a few passive components. The four resistors connected at Q3, Q4 and Q5 outputs will generate an 8-step graduation scale whose lower step is recognized as horizontal syncs by a TV set because Q6 resets the counter when it reaches the number '64'. In this way a reset pulse every 64μS will be created because 1MHz/64=15,625 Hz. This circuit, however, lacks vertical sync and we could see the greyscale image roll on our screen. As this is an experimental circuit, it can be used for test however. Some experts could extract a near 50 Hz signal by dividing the narrow reset pulse present at Q6 by 312 and mixing (enlarged to 160 μS by a monostable) to form a video signal to lock the image on a video display device's screen.

Ferranti produced another interesting chip named the *ZNA234* which makes available all the waveforms necessary to produce the crosshatch, dots, vertical and horizontal lines and greyscale test patterns on a television screen. The composite video output can be injected directly into the video input of a receiver or used to drive a video modulator for connection to the aerial socket. Just like its 'big brother', the ZNA134, this chip also generates a commutable-by-switch 525/625 line system although not fully broadcast compliant so that it cannot be employed in professional video equipments. In particular, a unique 2.5MHz crystal quartz is used for generating both 525-line and 625-line rasters which simply by changing the logic state at pin 2 enables you to switch between the two modes. The Irish *Donberg Electronics LTD* (www.dongerg.ie) and the British *Littlediode* (www.littlediode.com) can still furnish it, but it is quite expensive. In Fig. 29 there is a typical application note.

There are other methods for generating video synchronisms, but now you have possibilities outlined.

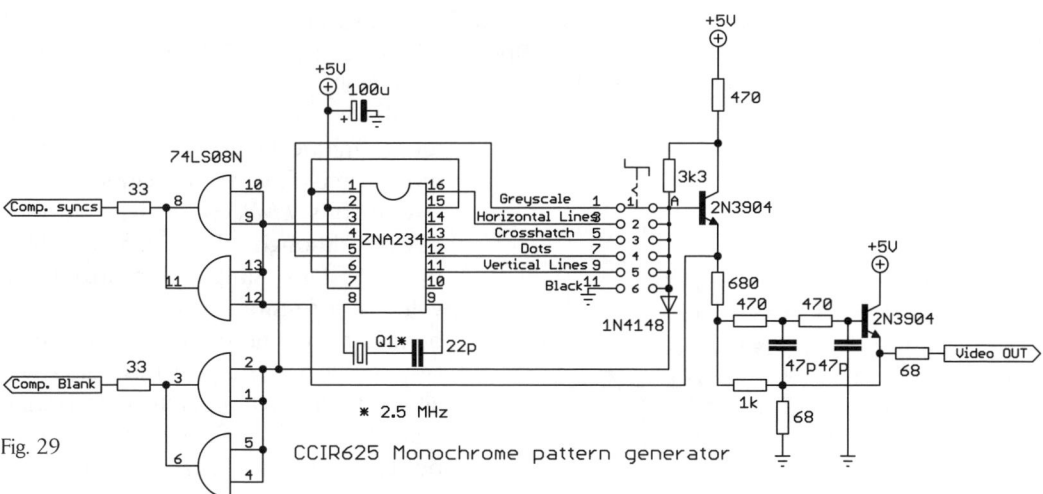

Fig. 29

2.5 Sync extraction

Now we will talk about the sync extraction from a composite video signal.

The simplest way is shown in Fig. 30: only three very cheap transistors and a few passive components!

From Q1 collector we get positive-going composite syncs (CS) at TTL level when the video signal has a 1Vpp level, while from Q2 and Q3 collectors we get, respectively, negative-going horizontal (HS) and vertical (VS) syncs always at TTL level. This happens because composite syncs pass through two different pass-band filters, each one centred on the frequency of the individual syncs.

Because HS and VS syncs are always present in a composite video signal, all the related signals needed for decoding colours and driving the display device, such as vertical and horizontal blanking, deflection signals, burst gate pulse and others, can be reconstructed starting from the syncs using simple analogue techniques including monostables.

However, the vertical deflection waveform obtained by analogue integration of broad pulses would result in sub-optimal revealing during field pulses due to thermal drift of the passive components constituting the vertical band-pass filter (or other factors). Now, because this integrated vertical negative signal must be used to trigger the vertical sync generator, in order to achieve a proper interlace, it is necessary for the field sync generator to be triggered after identical delays following the last picture line of both odd and even fields. If the field sync pulses occurred immediately following the end of the picture, then there would be a potential difference in triggering times since, at the end of the odd field, there would be only half a line duration following the line sync pulse before the start of the field sync pulses whereas at the end of the even field there would be a full line delay. Due to these factors, the two distinguished fields could slightly 'hop' up and down on screen, or in worst cases, corrupt interlace frames.

Over the years more elegant and improved sync separators have been designed for limiting interlace problems. One of them was composed of just a comparator with its threshold set around halfway between black level and the sync tips. The field sync pulses can be distinguished from the line syncs using simple analogue techniques. This was the method universally used until pseudo digital techniques were developed, which essentially took over.

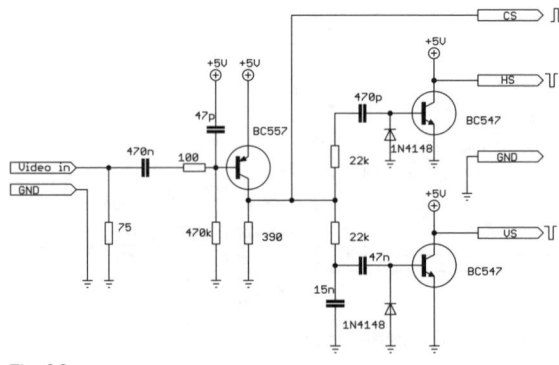

Fig. 30

One of the ICs that employs the latter technique (as well as having some other useful features) is National Semiconductor's *LM1881*, a most famous device. Although originally intended for NTSC systems, simply by changing the value of a resistor on pin R-SET, it can be also re-configured for use with PAL systems. In fact varying this value will allow the device to operate at horizontal scan rates up to 150 KHz. From a composite video with negative going syncs

2.5 Sync extraction

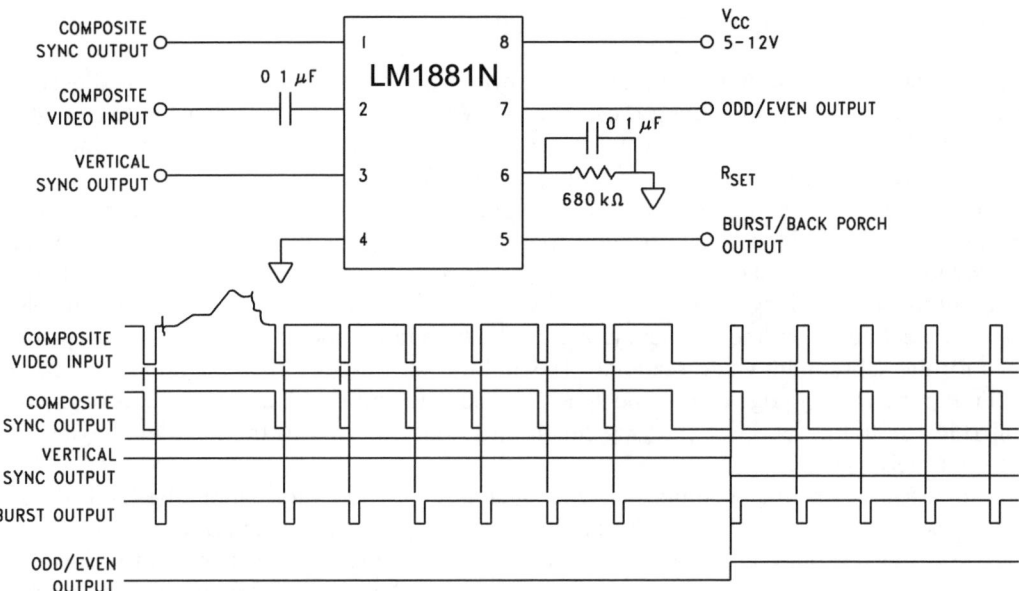

Fig. 31

ranging from 0.5 to 2 V peak-to-peak, the following signal outputs are extracted: Composite Sync, Vertical Sync, Odd/Even field pulse (a 25 Hz square wave) and a Burst/Back Porch output pulse. A very useful feature of the device is a default triggered vertical sync output when it is being fed with non-standard signals, such as from home computers or consoles, some of which do not produce the standard normal vertical sync pulse. The pin connections and block diagram are shown in Fig. 31 where you can see the only other components required are an input capacitor and an R-SET decoupling capacitor, both of which are recommended to be 0.1µF polyester. The device operates from a single supply between +5 and +12 V and the outputs are at CMOS level.

The composite sync output is a reproduction of the input signal waveform with all the information above black level removed. Normally the signal source for the LM1881 is assumed to be clean and relatively noise-free, but some sources may have excessive video peaking, causing high frequency video and chroma components to extend below the black level reference. A clean composite sync signal can be generated from such sources by filtering the input using a series 680 Ω resistor and a 470 pF capacitor to ground for 75 Ω source impedances which will form a low pass filter with a corner frequency of 500 kHz. However this filter can cause a little shift delay of between 40 ns up to 200 ns due. This delay will usually be irrelevant but beware it could contribute to the sync delay produced by any additional signal processing.

The vertical sync output is derived by internally integrating the composite waveform. Due to the long duty cycle, the serrated vertical sync pulses are able to charge an internal capacitor at a fixed threshold. Once this threshold has been reached, the next serration pulse triggers a R/S flip-flop and initiates the vertical output pulse. An internal counter is started which, upon reaching eight, resets the circuit and terminates the pulse. If the incoming vertical sync is not serrated, then the capacitor is allowed to charge to a second threshold which automatically starts the output sequence.

The Odd/Even field pulse can be useful in frame memory storage applications. The pulse is derived by further integration of the composite sync waveform. A capacitor is charged between sync pulses and discharged during them, the result of which is fed to a logic network which compares the output of a flip-flop with the vertical sync. The resultant 25 Hz square wave pulse is low during the even fields and high during the odd fields.

The Burst/Back Porch output pulse can be utilised to either retrieve the chrominance burst from the composite video signal (for subcarrier synchronising) or as a clamp for DC restoration of a video waveform. In a composite video signal the chroma burst is located on the backporch of the horizontal blanking period, as we will see in the next chapter, which is approximately 5 µS wide, depending on line frequency. The Back Porch pulse also acts as the black level reference for the subsequent video scan line. This pulse is obtained by charging another capacitor starting at the trailing edge of the line sync pulses at the end of a line. The output of pin 5 is pulled low until the capacitor charging circuit times out 4µS later as the next line begins, and then returned high.

Apart from extracting a composite sync signal free of video information, the LM1881 outputs allow some interesting applications to be developed. As mentioned above, the burst gate/back porch clamp pulse allows DC restoration of the original video waveform for display or re-modulation onto an RF carrier, and retrieval of the colour burst for colour synchronisation and decoding into RGB components. For frame memory storage applications, the odd/even field square wave pulse allows identification of the appropriate field ensuring the correct read or write sequence field-by-field. The vertical pulse output is useful since it begins at a precise time: the rising edge of the first vertical serration pulse in the sync waveform. This means that individual lines within the vertical blanking period (or anywhere within the active scan line period) can easily be extracted by counting the required number of transitions in the composite sync waveform following the start of the vertical output pulse. An example of a single line selector is shown in Fig. 32.

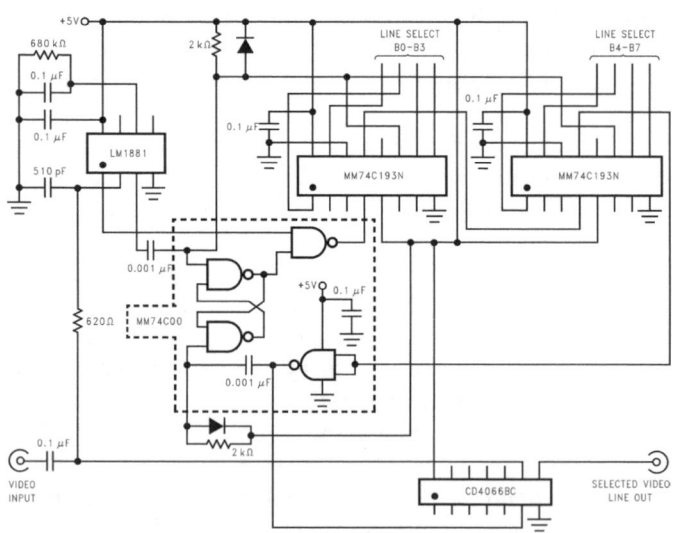

Fig. 32

The LM1881 is still well distributed in both DIL and SMD version from many large distributors even, such as Farnell and RS Components. It will be employed in some future projects presented in this book.

Elantec (now Intersil) some years later introduced a more sophisticated version of the LM1881, called the *EL4581*. Now although this last chip is pin-to-pin compatible with LM1881, its absolute maximum voltage supply is +7V meaning it cannot be substituted in those circuits where the DC supply is beyond this.

2.5 Sync extraction

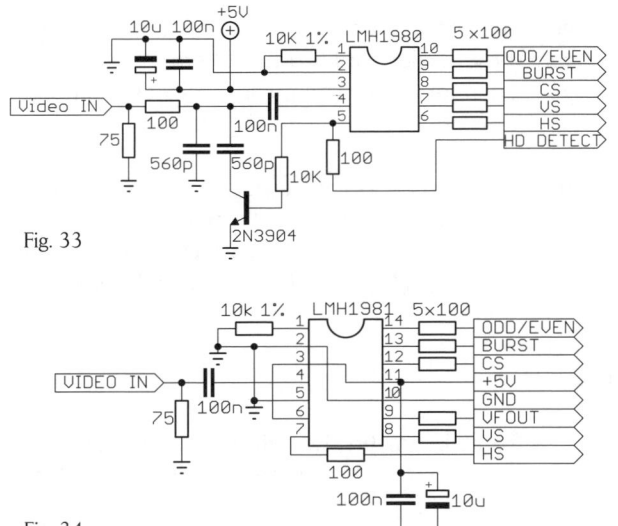

Fig. 33

Fig. 34

Where possible, National encourages migration from the LM1881 to the *LMH1980*, a new *Standard & High Definition/Personal Computer* (SD/HD/PC) video sync separator device designed to replace the old LM1881 sync separator in consumer, security and surveillance, and additionally industrial video applications where high-definition compatibility is needed (Fig. 33). In broadcast and professional video systems where timing jitter is a critical system parameter, National designed yet another sync separator, the *LMH1981*, which offers the best jitter performance while supporting all SD & HD analogue video standards (Fig. 34). Both LMH1980 and LMH1981 are packaged as 14-Pin TSSOP whose dimensions are 3x3 millimetres making it is almost impossible to solder by hand!

In 1995, Gennum Corporation introduced some LM1881 equivalent chips: the *GS1881*, *GS4881* and *GS4981*. The GS1881 is fully compatible with the LM1881; GS4881 is identical to the GS1881 but features a noise-immune back porch pulse which maintains a constant horizontal rate during the vertical interval. The GS4981 is identical to the GS4881, except that it provides horizontal sync in place of the odd/even output. Gennum also introduced, in 2004, another perfect, improved and precision (and more costly) replacement for the LM1881, called the *GS4882*. Another chip compatible to LM1881 was the *GS4982* with a little difference on

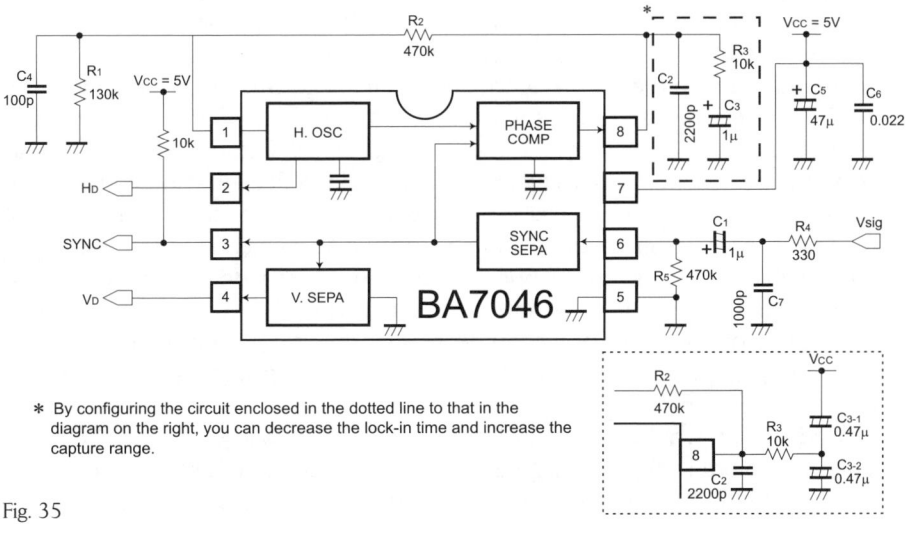

* By configuring the circuit enclosed in the dotted line to that in the diagram on the right, you can decrease the lock-in time and increase the capture range.

Fig. 35

pin 1, in that instead of composite syncs, you get a horizontal sync output. Gennum ICs are hard to find lately; searching on the internet probably some dealer could furnish them.

Rohm Semiconductors produces some sync separator chips. The most readily available on the market is the *BA7046,* an 8 pin DIP device distributed by RS Components. In Fig. 35 is described an application example.

Many other sync separator chips were designed and produced especially for B&W and colour TV sets, video recorders and TV monitors among the most popular are the *TDA2579 & TDA2595* (Philips), *MC44145* (Motorola), *SDA 9257* (Siemens), *ZXFV4583* (Zetex) and *TDA8181* (ST Microelectronics), based essentially on PLL (*Phase Locked Loop*) technology to extract video synchronisms.

2.6 A curious television scanning alternative!

In a line scanning system the exploring electronic beam (in the camera), after each line, should make a blank path, the horizontal blanking, to begin analyzing the next line, equivalent to 12 µS. As each line is 64µS, removing the 12 µs leaves 52 µs which amounts to only 81.25% of the total time that is actually used for each line.

Only 575 of the 625 lines are actually utilized you will recall and the total time to transmit only the visual information of those lines is: $575*52*10^{-6}=0.299$ seconds, i.e. 29.9 ms. Since 625 lines compose an entire frame whose length is twice a 20 ms single field timing, we can say that in a frame we get 2*20/29.9*100=74.75% of total time in which the electronic beam scans or recomposes a complete picture. The remaining time is 'lost' in sync pulses, switching off the beam and other informations not directly visible. What a waste, you may think!

At the beginning of the '50s in the French *Laboratoires R. Derveaux* was designed and patented (patent #GB750187) a spiral television exploring technique which offered many evident advantages, so that at the beginning it was taken in consideration for the industrial and military fields. Its adoption in the public television service, however, found serious obstacles in the wide spreading of line scanning with which all countries had already set up their TV services.

But with this solution, the exploring beam scanned in 1/25 of a second the entire image along a regular spiral line proceeding from the centre to the edge or vice versa (as in a vinyl record) and then instantly back to the centre for the following image analysis. Each point of the image modulated the thin ray of electrons, as was the case of lines, but with a different path and with greater continuity, rather than by successive straight lines. The exploring spiral was obtained in this case by means of an increasing and rotating field deflection. At the end of the exploration transit, it was sufficient just to clear the field deflection so that the contact point of the beam with the image, or electronic spot, would be returned to the centre of the screen. Of course, it could be possible to proceed in the opposite direction, with a downward spiral from the periphery to the centre, or even have an intertwined network of lines for image analysis, with a combination of two or more increasing and decreasing spirals. Or for a colour television could be assigned a unique spiral for each fundamental colour transmitted sequentially and interspaced by a sync colour signal.

Another major advantage would be the lossless time during analysis of images. With this process only after a full exploration of the image did they exist a need to return to the centre in order to begin a new spiral. So the time lost in this case would not exceed 2-3% of total time of

2.6 A curious television scanning alternative!

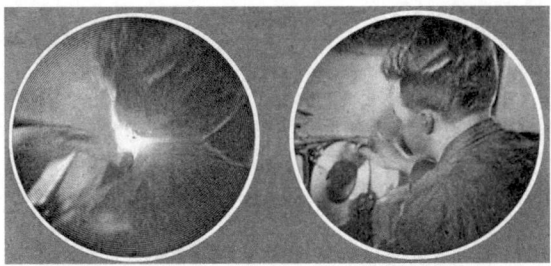

Fig. 36 - Left: exploration spirals appear in picture; right: the same spirals became invisible after calibration.

analysis rather than 25% of the line scanning. Indeed, with the adoption of both increasing and decreasing spirals, that blank time could even have been abolished.

So spiral scanning fully exploited the bandwidth frequency allocated for transmission. Also, outside of the so-called voltage driver, which provided the right correlation of plot analysis for departing and arriving spirals, there was no need any other sync signals, whereas in the traditional system these signals occupy a significant place.

Overall we can say that, thanks to its unique features, this new procedure had presented a very favourable technical solution. Even the transmitter power was better used thus to the benefit of weaker signals.

However, all these properties, despite forming a significant advantage in favour of the exploration spiral scanning, could not supplant the predominant line system since it involved huge financial interests in consequence of the high number of working equipment and under construction.

Like many other brilliant ideas, spiral video exploration scanning system remained in the experimental phase or was used in specific industry sectors such as the remote monitoring of hazardous jobs, handling of radioactive substances, etc. And finally the spiral exploration system, being constantly perpendicular to the radius of the cathode ray tube, is also well suited to the retransmission of images that appear on radar screens (Fig. 36).

---

[1]. A. Meucci was recognized as the first inventor of the telephone by the *United States House of Representatives* in *House Resolution 269* dated June 11th, 2002. The resolution states that *"If Meucci had been able to pay the $10 fee to maintain the caveat after 1874, no patent could have been issued to Bell."* However, this declaration is non-binding and has no legal effect.

[2]. L. Le Prince is considered by many film historians as the true father of motion pictures who shot the first moving pictures on paper film using a single lens camera in 1888, several years before Edison and Lumière brothers. He was never able to perform a planned public demonstration in the United States because he mysteriously disappeared from a train in September 16th, 1890.

3. Colour Standard Systems

Prior to a technical explanation of how we arrived at a backward compatible colour television standard, we must highlight our ultimate visual receiver organ, the **EYE**, with a brief description.

3.1 The Human Eye

The eye is a very sophisticated optical system, the most complex known, providing images of real objects that are in any spatial position.

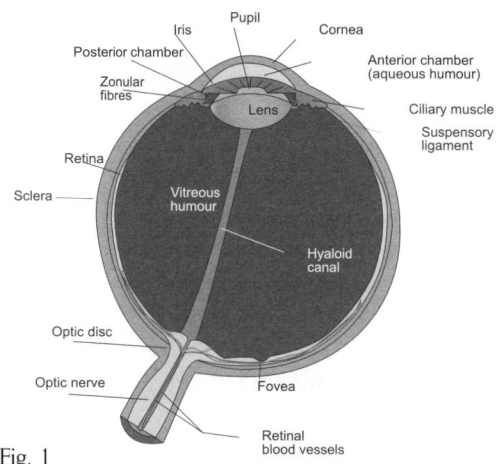

Fig. 1

The elements that constitute the optical eye are: the *cornea*, from which ray lights enter; the *aqueous humour*, a liquid substance refracting rays like if it were water; the *eye lens*, an element with the shape of a biconvex lens; the *vitreous humour*, another liquid substance placed between the lens and the *retina*, the light sensitive surface on which the image is formed (Fig. 1).

The principle of operation is largely similar to a photographic camera with a biconvex lens. If an object is placed at great distance from the biconvex lens or, in extreme, at infinite distance, its image is formed at a distance from the lens equal to that of the lens itself, i.e. it is formed precisely on its focal plane.

If the object is placed at a predetermined distance from the lens, its image is formed behind the focal plane. In the case of the camera, the plane, which forms the image of any object, must necessarily coincide with the plane of the photographic film.

Since the camera objective has a fixed focal length (depending on the characteristics of construction), to be possible that the image is always formed in the desired position, whatever the distance from the object, you have to move the lens forward and backward in relation to the target plane of the film. This operation is commonly called *focus*.

The image on the eye is different from that one given by the camera lens, first because it is curved. Indeed, it must be formed on the retina that is curved because it covers the internal back of the spherical eyeball.

All these resources contribute to the formation of the image, but the focus is made by the eye's lens, contracting and expanding so to vary the focal length. This operation is called *accommodation*.

When the objects are at infinity distance, the eye is not accommodating, and when instead it has to watch neighbours items it must accommodate, as for example when we read. Another similarity between the eye and the camera is the presence of the *iris*, which, like the aperture of the cameras, by restricting or enlarging it, adjusts the luminous flux that has passed through.

The hole delimited by the iris is called the *pupil*: it expands when the lighting is weak and restricts when it is very luminous.

The retina is a nervous tissue in which there are components sensitive to light, the photo-receptor cells which are of two types: *rod cells* and *cone cells* (or simply *rods & cones*), so called because of their characteristic shape.

The cones, which are used in daytime vision, are combined in an eye's area called *central fovea*. They thin out moving away from this area and reach a number of about 8-10 million for each eye. The cones are of three types and are different for the three different pigments that absorb in different ways the three main parts, red, green and blue, of spectrum light. These pigments are generally called *iodopsin*.

Outside the fovea there are about 120 millions of rods that are much more sensitive than cones and are able to reveal even a single photon. Rods cannot distinguish colours, but are responsible for low-light black & white vision; they work well in dim light as they contain a pigment (the *rhodopsin*, consisted mainly by a protein, the *opsin*), visual purple, which is sensitive at low light intensity, but saturates at higher intensities.

Cones and rods are the terminations of the optic nerve filaments. The point where these are inserted in the eyeball does not have any light sensitivity and therefore is called the *blind spot*.

The fovea is located at the optical axis with the eye retina. By night if we look directly at a little shining star, we cannot see it: its image is formed, in fact, on the fovea, where there are no rods. We instead can watch pointing it slightly aslant.

The light induces chemical modifications on rhodopsin and iodopsin producing an electrical stimulus that through various processes will reach the *brain's visual cortex*, where, further elaborated, finally is created the phenomenon of vision.

An important feature of retina is the retaining of images for around 90 ms after their disappearances. This phenomenon is called the *Persistence of vision* which was brought to the development of cinematography and television technology, where a rated sequence of still frames creates the illusion of movements.

3.2 From eye to its electronic representation

Television in black and white (B&W TV) was just one stage of the unfulfilled journey to the most remote lands of the colour television (TVC).

To make the journey more assailable research has contributed a physiological principle of the eye's nature where a full colour vision can be given by the appropriate composition of just three fundamental colours: Red, Green and Blue (the so-called *trichromacy* or *trichromaticism*).

In fact, it was found experimentally that the response of the human eye to colours can be represented by the integration of sensations from three basic types of photoreceptor 'cone' cells, and each type responds differently to light radiations. The type 'L' (Long) responds most to long wavelength light (λ), usually 610 nm (nanometres), peaking in the yellow-red region; the type 'M' (Medium) to medium-wavelength light, usually 535 nm, peaking at green and the type 'S' (Short) to short-wavelength light of a blue-violet colour, usually 470 nm (Fig. 2).

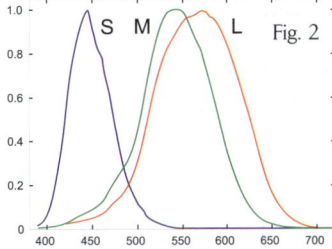
Fig. 2

When we watch a coloured object, the cones are stimulated at various levels and via the brain's integration of the

Chapter 3. Colour Standard Systems

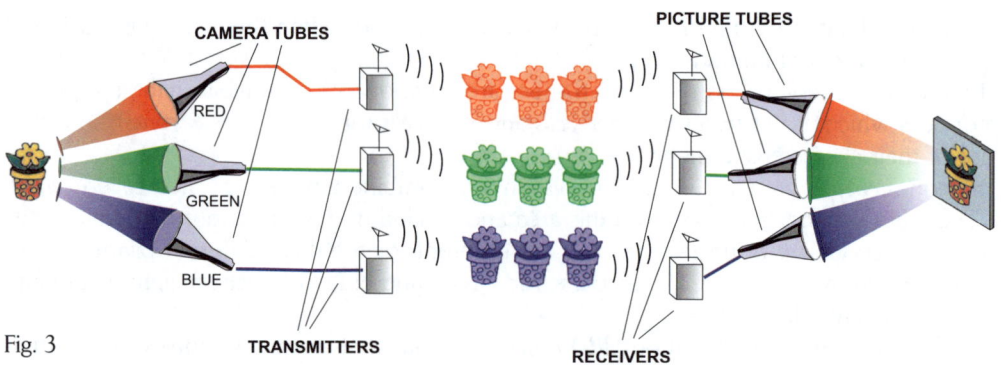

Fig. 3

three stimuli will give the feeling of the resulting colour.

So if a TVC system is capable of analyzing the content of the three fundamental colours and then the equivalent electrical signals are proportional to the requisite amount, the captured images will be certainly considered near identical to the original ones.

This policy found its immediate application in a system which provided, in transmission, three different camera tubes, each one equipped with a colour filter separately for each of the three colours The tubes would simultaneously focus the same image of the subject under observation through optical arrangements.

At the receiver, the resulting signals, each channelled via a separate carrier wave for each colour to activate, would produce three kinescope images, each one being sensitive to its primary colour; the overlay of the three partial images through optical systems leading to the resulting colour image (Fig. 3).

Such a simultaneous system, even if it had various beneficial effects, impacted against the technical difficulties of optical and electrical nature, as well as a considerable economical effort, which killed adoption.

A possible system, known as *Colour Field Sequential System*, was proposed by *Peter Carl Goldmark* from CBS and was indeed accepted straightaway by Federal Communications Commission (FCC), the competent authorities of the United States.

We know that television is a sequential series of still frames, following each other quickly enough to avoid flickering. The idea was to extend this technique to colour reproduction using a tricolour filter wheel revolving in front of a pickup tube at frame rate so that stills of each primary colour components of the television scene would be sent in rapid succession, thus rendered as a consecutive series of three groups of signals corresponding neatly to red, green and blue colours.

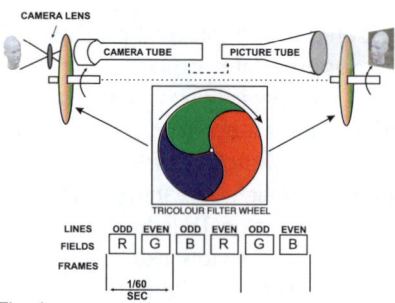

Fig. 4

If in the receiver a similar wheel span synchronously in front of a normal type of kinescope, the merging of three images, derived from the usual phenomenon of persistence on the retina, would give the resulting coloured image (Fig. 4).

However the main problem here was the extreme flickering. Because the repetition rate of each primary colour was now only one-third of the just acceptable TV frame repetition fre-

3.2 From eye to its electronic representation

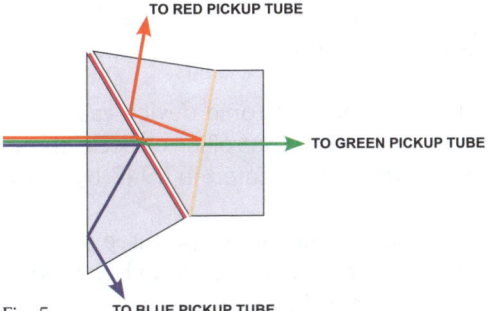

Fig. 5

quency, coloured objects flickered disquietingly; even if a kind of storage were invented to neutralize this, moving coloured objects would cross the screen with a series of blurred and jumping pictures. The trouble was solved by triplication of the frame rate; yet this system (like the previous simultaneous system) had the major drawback to bind a triple-wide video & transmission bandwidth and therefore it could not be received by B&W television sets.

A substantial compression of the characteristics represented by the relative reduction in the number of lines, in the frame frequency and the horizontal definition, allowed a return to normal standard bandwidth but with the collateral effects of a lower visual quality and (above all) a flicker deterioration, here especially evident for this succession of colours. But in spite of these adjustments, it was incompatible to then normal black & white TV receivers.

So some researchers, determined and tenacious, focused their efforts on a system that, within the regulatory standard, allowed reaching the same quality of black and white television for 'normal' receivers. Several systems had been studied, some of which had passed the experimental stage to be used commercially on a large scale.

A valuable assistance was also provided in this case from a characteristic of the human eye as a result of its reduced sensitivity to the perception of colour changes, i.e. roughly half that of light intensity variations, which could translate in requiring less colour visual information. The transmission bandwidth to be used could be restricted accordingly.

One of these systems was proposed by *David Sarnoff* from RCA, which, taking advantage of the eye's characteristics, employed for the transmission three normal camera tubes; the partial images for the three fundamental colours were focused by the objective lens on related targets by means of suitable optical colour separators, such as the *Dichroic mirrors* (Fig. 5) or coloured *filters*. The moving images were analyzed at standard scan rate.

The three complete groups of signals were subsequently treated to compress them into 4 MHz (for the American system), the total bandwidth containing all the information necessary for colour reproduction of the images.

Thus each of the three signals after having passed through a *Low Pass Filter* entered into an electronic device called an *Encoder*, which provided the appropriate combination of signals.

The signal to be transmitted from the resulting combination was a complex signal, which contained a portion corresponding to the sum of the three Red, Green and Blue signals which

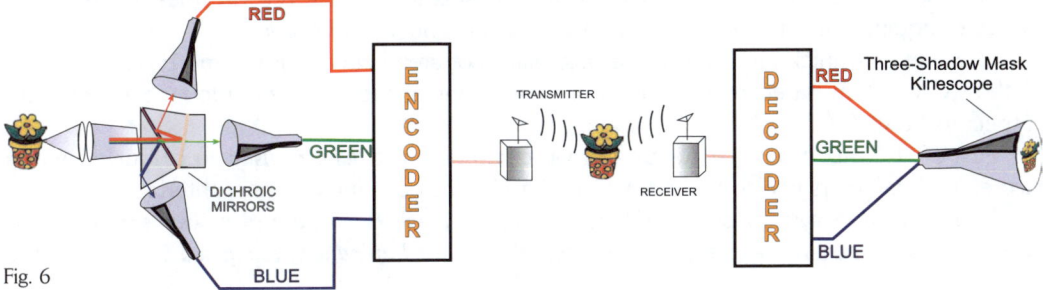

Fig. 6

Chapter 3. Colour Standard Systems

represented the light intensity sufficient by itself for the correct reproduction on B&W TV sets. The remaining part of the three signal components represented the individual fundamental colours and their relative 'strength'.

In the receiver the incoming signal is demodulated via an electronic device called the *Decoder* and then separated into its three colour components before controlling a special three-gun cathode kinescope called *Three-Gun Shadow Mask Kinescope*, patented in 1947 by *Alfred Schroeder*, a retired RCA researcher (Fig. 6).

A shadow mask is a screen full of tiny perforations. In a single kinescope three separate electron guns, one for each colour were combined. The electron guns shot through tiny holes in the shadow mask, each hitting dots of red, green, or blue phosphors.

Control of the three beams was applied in common to the three electron emitter cathodes of the tube to form the resulting signal, which provided the light intensity of the individual image points. In addition at the same time, one of the three partial signals which provided the colour content related to the light intensity of image, was applied to each of the cathodes themselves.

In early prototypes, there were 150,000 clusters of phosphors, each one containing all three primary colours. Each gun had to hit the right phosphor via a tremendous task of alignment. Everyone said it could not be done, but it worked!

RCA's first shadow mask colour televisions needed much more tuning than modern colour televisions do, but in this way, by eliminating mirrors, multiple CRTs, rotating wheels, synchronized motors, etc., all the television chain was facilitated and streamlined.

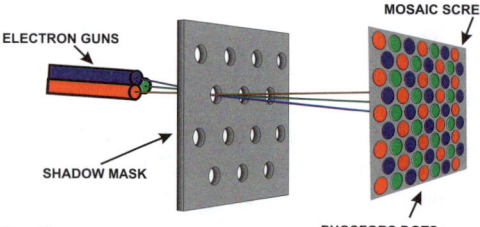

Fig. 7

Figure 7 clarifies how this very clever mechanism can result in a colour image that is an exact reproduction of the original.

A receiver equipped with this new kinescope was equally suitable for the reception of television in black and white because in the latter case, the colour signal would be identical for the three beams thus producing a grey scale image.

Of course, an alternative could have been that instead of a single kinescope with three beams, three single-beam kinescopes could be employed and via colour decoding driven by the three fundamental colours, so you could get three pictures which would be composed and aligned thereafter using optics.

This latter solution, although simpler, was very cumbersome and then not practical in normal domestic televisions but it has had a wide spread in more recent times for television projector equipment on projection flat screen such as the *Video Projector*.

The previous quick hints about the possibilities of creating colour TV and broader aspects of the RCA system raised more general considerations of the needs of colour television with respect to B&W TV.

First of all, while B&W TV needs a single parameter, the *light intensity*, in colour television two extra parameters had to be implemented and selected using different criteria.

So in the RCA system previously described, for black & white and colour simultaneous combining, the chosen parameters were *light intensity* (or *luminance*), *colour tint* (or *hue*, i.e. the spectral characteristic expressed by the light wavelength in the range of the visible spectrum) and *saturation* (describing the amount by which the colour would be diluted by white colour to get the fundamental colour component of each element picture) whilst in the CBS sequential system a different list of parameters was chosen consisting of the individual amount of each of the three fundamental colours that, combined in additive synthesis, gave rise to the perception of the original colour.

It is appropriate to point out that the electrical signals are of course all of the same kind and type, for any colour they match.

They are the values of the features they represent and their relative sequence in the composition of the total signal. The only prerequisite to ensure that the selection of colour made by optical filters for the transmission will be equal in the receiver is achieved by similar filters or corresponding dye fluorescent material both in the capturing and reproduction equipment.

But what elements could be used to transmit the colour information?

Clearly on only the elements involved in the television mechanism i.e. on points or lines or on the frames & video frequencies. The transmission carrier wave does not enter into this case (apart from the previous system, rejected a priori, of three contemporary carrier waves) because, once exhausted its function as a vehicle, it is eliminated.

So RCA and CBS system, respectively, were based on line and frame interlacing; a third system developed by *CTI* (*Colour Television Incorporated*), likewise the other two contenders, implemented an interlacing of lines for each frame being composed of a succession of rows in alternating colours.

RCA's final system, which corresponded to a criterion of standard line and frame frequencies, included the signals for the colour informations in the spaces free in the frequency spectrum components of the normal video signal.

These signals were 'superimposed' on a subcarrier wave frequency suitably chosen in this spectrum; the total signal was then ready to modulate the video transmission carrier wave, in the usual manner.

This is the manner by which the NTSC colour system of *Hazeltine Corporation* worked, which also developed and licensed many of the basic concepts of the NTSC colour television system. It interposed the informations of the three fundamental colours in the spectrum of the basic signal normally corresponding to light intensity.

The RCA system owned the unusual feature, compared to other systems at that time, to implement the simultaneous transmission of all information, without visibly lowering the quality of the normal black and white standard.

The colour information gave essentially no trouble to the reception in black and white on 'old' TV receivers, because in this case the receiving colour signals were removed automatically.

Of course, each system had limitations of operation mainly related to the radio frequency bandwidth of the channel assigned to the normal TV broadcast.

So the CBS system, which had a significant flicker visible in individual monochrome frames, needed an increase in the frequency frame at the expense of a lower horizontal definition, however not perceived by the eye for the compensating effect of colour.

In the CTI system the flicker increment was significant because the normal frequency of 30 fps (frames per second) was reduced to 10 fps for each colour and this system was soon discarded.

The RCA system instead ensured an increased level of detail because the items are repeated at the same black & white normal rate including the colour information without worsening the flickering.

The RCA colour stable system no doubt opened the doors to the wider development of fully electronic colour television.

The high costs for modifying the existing two million monochrome receivers to follow the CBS colour field-sequential scanning system were a major factor in the CBS debacle. If adopted, a consumer would have overspent at least $25 for adapting a CBS colour wheel inside each set which, for those times, corresponded to about $250 nowadays. In addition, those who had not yet modified their monochrome receivers, would also have had no opportunity to follow the programs scheduled by television companies which adopted the new backward incompatible television colour system. So the CBS colour system, after just a few years of life, was abandoned in 1954. Ironically, in the United States cameras using the CBS colour wheel system continued to be used for scientific research for several more decades, including at least one of the NASA moon landings in the '70s!

3.3 NTSC & PAL Encoding

Let's see in detail the clever idea that allowed the 'miracle' of colour transmission backward compatibility. NTSC & PAL systems are quite similar in philosophy of concepts and therefore will be treated together.

It is generally sufficient for a capture device, such as a camera tube, to generate greyscale images which, added to composite synchronisms, create a B&W 'complete' video. *Hazeltine Corporation* and RCA's engineers faced many challenges and troubles to solve before creating a 'colour' video signal that on a 'normal' B&W TV set would not display pictures with false grey gradations and yet on a colour TV set would not cause misleading colours and wrong greys.

Valuable aids were the studies and experiments completed by scientists as *Isaac Newton* and *Hermann Grassmann*.

Newton, especially by pointing a ray of sunlight (white light) towards the side of a glass prism, had seen the breakdown of white light into seven main coloured rays: red, orange, yellow, green, blue, indigo and violet. However subsequent observations showed that rainbow-coloured rays were not composed of the seven familiar colours, but that each one 'overshot' with the appearance of numerous intermediate shades in transitions between colours. As previously stated, the human eye's retina, however, distinguishes only three main zones: a blue, a green and a red zone within the visible band ranging from a wavelength of about 380 to 780 nm via a large number of light-sensitive cells (rods & cones).

In order to create three distinct video signals, each one corresponding to the eye's main colour sensitivity, we should use three monochrome capture devices which through red, green and blue filters, one for each device, would generate three distinct monochromatic video components called *RGB Components*.

Next step was to sum these components someway to obtain a valid greyscale image comparable to that generated by a single monochrome capture device without optical filtering, for backward compatibility transmission. The initial idea was to divide the RGB Components in equal percentage and then mix them to produce a greyscale image but the result was a very dark image! The issue was due to the human eye's characteristic and its behaviour with light.

In particular, because only cones peak into three specific regions of visible spectrum cor-

3.3 NTSC & PAL Encoding

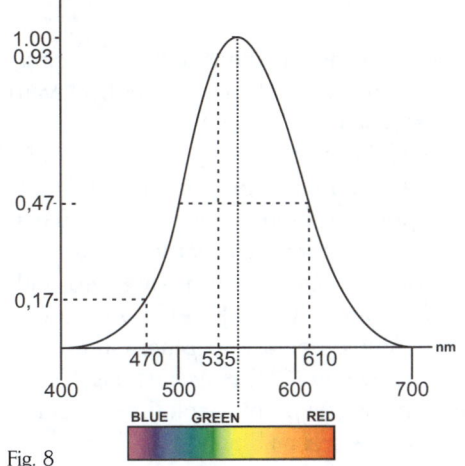

Fig. 8

responding to <u>R</u>ed, <u>G</u>reen and <u>B</u>lue (for this called *Primary Colours*, often referred simply as *RGB Colours*), they also do not respond linearly to rainbow-coloured light, peaking at 560nm (yellow zone) of radiance spectrum, so that all saturated blue and red colours appear quite dark and all saturated yellows are quite light.

The graph in Fig. 8 shows that the relative value (or coefficient) of red is 0.47, of green 0.93 and of blue 0.17, with the maximum value, i.e. 1, at yellow, according to the law of *Hermann Grassmann* who first discovered it.

Instead rods are numerically twenty times more prevalent than cones and are very much more sensitive to the brightness content of an image, in television called *Luminance* or *Luma*, shortly Y. This is the reason for which fine details of nice pictures are perceived in black and white only and also why under very weak illumination we see in grey shades.

In colour television, however, we should only play with a 'RGB triad' and from it we should generate both valid greyscale images and coloured images.

In a B&W video signal, 0% luminance corresponds to black level and 100% to white level. In the middle there are all the grey tones.

Now, according to Newton's discoveries, by mixing all primary colours between them, we get the white colour, i.e. White=Red+Green+Blue.

Since the luminance is a 'RBG mixture', it can therefore be deduced that the eye's maximum sensitivity at white, summing all the red, green and blue coefficients according to Grassmann's law, is:

0.47+0.93+0.17=1.57.

Dividing all the terms by 1.57, the following expression is created respect to unity:

(0.47/1.57)+(0.93/1.57)+(0.17/1.57)=1

or simplifying:

0.30 + 0.59 + 0.11 = 1.

Since 100% of luminance Y corresponds to white (i.e. R+G+B), then:

Y = 30% Red + 59% Green + 11% Blue.

This is the reason why equal proportions of red, green and blue are not combined to produce the luminance signal! Furthermore such a luminance signal so created from the weighted RGB components was already sufficient to be received untroubled by a traditional TV set displaying so called 'panchromatic greyscale' images quite similar to old B&W television system.

Chapter 3. Colour Standard Systems

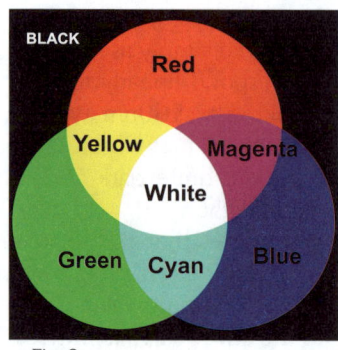

Fig. 9

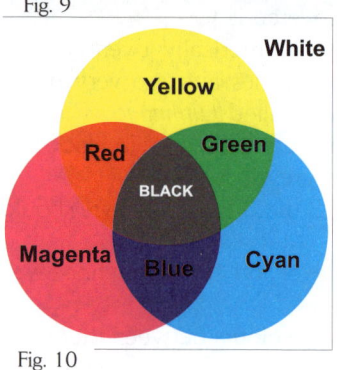

Fig. 10

Now it is the time to introduce the *Additive & Subtractive Mixing* concepts to continue in explaining about backward compatibility in colour television.

We know that all the visible colours are perceived by mixing the primary colours with different proportions. However combining the saturated primary colours one-by-one between them we obtain the so called *Complementary Colours*. For example, the *Yellow* colour is obtained by mixing Red with Green, *Magenta*=Red+Blue and *Cyan*=Blue+Green; Yellow, Magenta and Cyan are complementary to Blue, Green and Red, respectively. This is known as *Additive Mixing* (Fig. 9).

As previously stated, mixing all primary colours we get the white colour, i.e. White=Red+Green+Blue.

However there is another method for generating the white colour, i.e. by adding a primary colour to its complementary. For example if Blue is added to its complementary Yellow, then since Yellow is Red plus Green, we get: Yellow + Blue = (Red+Green) + Blue = White. Complementary colours can also be produced by a process called *Subtractive Mixing*. Yellow, for example, can be produced by subtracting Blue from White. Since W=R+G+B, then W-B=(R+G+B)-B=R+G=Yellow. Similarly, W-R-G-B=black (no colour), W-R=G+B=Cyan, W-G=R+B=Magenta (Fig. 10).

Having said this and laid the first basis for a backward compatible colour television system, it remained the problem how to add the RGB signals to luminance without causing interference.

The genial idea was to introduce a subcarrier modulated in phase and amplitude quadrature by red and blue signals from which the luminance was subtracted. These were called *Colour Difference Signals*, named briefly *R-Y* and *B-Y* signals. The reason behind the choice of utilizing only two colours in collaboration with the luminance for creating a colour video signal is very simple. Since the colour information is effectively coded in the luminance signal, it is only necessary to transmit two further signals in order to be able to obtain the separate R, G and B signals in the receiver or, in other words, it suffices that only two colour difference signals require to be transmitted. Furthermore, although any pair of difference signal can be used, it was seen that because the Y component already contains a large part of the green information from the image, i.e. 59%, G-Y becomes redundant. In addition, the relatively small G-Y signal would be more vulnerable to noise in the transmission system than the larger B-Y and R-Y signals and therefore it was discharged. The RGB components were recovered in the receiver in this way: Red=(R-Y)+Y, Blue=(B-Y)+Y and since Y=R+G+B, Green=Y-R-B.

Before driving the subcarrier modulators, B-Y and R-Y were scaled in the percentage of 49.5% and 87.7% respectively, numbers computed using a complex matrix calculus in order to maintain a slightly attenuated level of subcarrier-modulated 'colour' signal within an appropriate range of values for the ultimate video signal. Indeed otherwise by adding the luminance signal to the modulated subcarrier, it could sometimes actually exceed the level of white or even be below the level of syncs. That would become really serious because the modulated subcarrier signal can affect the sync separator and the deflection circuits of the receiving device. More-

3.3 NTSC & PAL Encoding

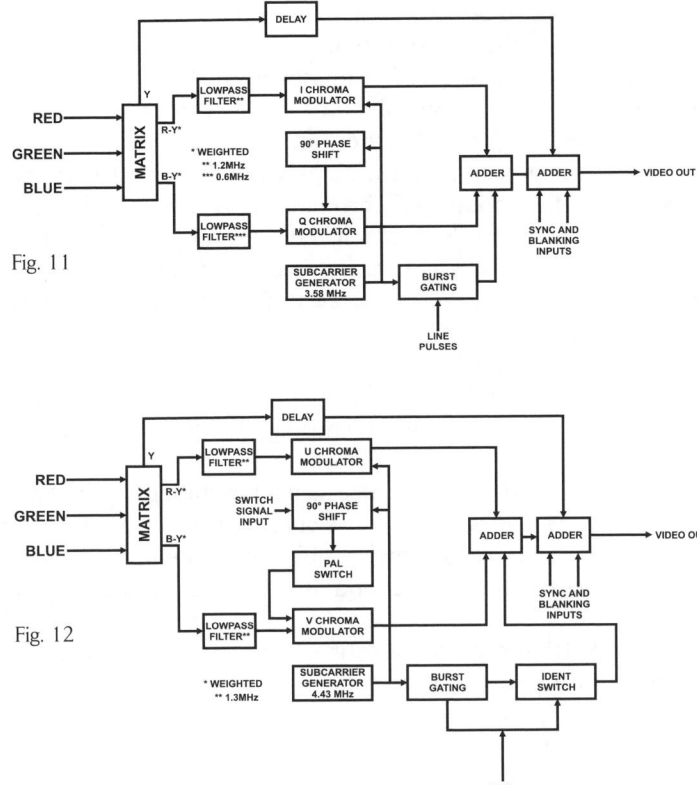

Fig. 11

Fig. 12

over, whereas the ultimate video signal is eventually transmitted using RF amplitude modulation, it is important to ensure that levels are maintained within the limits of possible overmodulation without affecting signal reception. Consequently, these coefficients were determined for the levels of amplitude subcarrier signal regardless of transmitted colour within the ranges in which they can guarantee a correct decoding of the signal.

The scaled versions of B-Y & R-Y were briefly renamed *I & Q* for NTSC (which stands for *In Phase* and *Quadrature Phase*) and *U & V* for PAL (which stands for *Unvarying* and *Varying*). In addition both components were low pass filtered (0.6 MHz for Q and 1.3 MHz for I for NTSC; 1.3 MHz for both PAL components) for further restricting the total video bandwidth (Fig. 11 & 12).

I & Q (or U & V for PAL) are now ready to feed two special types of modulators called *Balanced Modulators* which give automatic subcarrier suppression when there is no modulating signal tied to them. When a signal is sent to a balanced modulator, its output is proportional to the level of the signal being fed, with one special attribute: if the signal is at its maximum positive peak, the modulator's subcarrier output is at 0 degrees phase, but if it is at its minimum negative peak, the output is at 180 degrees phase.

These two balanced modulators are driven one by I (U for PAL) and the other one by Q (V for PAL).

To prevent these mixed modulators from interfering with each other, we still have one more trick to perform, i.e. to shift one of the modulators' outputs by 90 degrees with respect to the other. So, the I (or U) modulator will vary in amplitude, and, when notified to do so by I (or U) input signal, also vary in phase flipping between 0 and 180 degrees. The Q (or V) modulator, on the other hand, also will vary in amplitude but will vary in phase between 90 and 270 degrees.

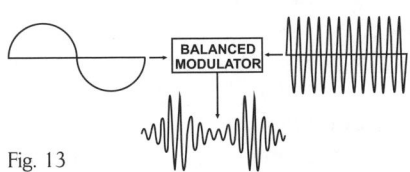

Fig. 13

Finally these two modulators' outputs are buffered and mixed together to create the so-called *Chrominance*

55

Chapter 3. Colour Standard Systems

or *Chroma* signal, shortly *C*. Depending on what colour is present in the input signal, you will get a particular phase of the subcarrier at that point that is a sum of the two balanced modulators' mix, most of the time somewhere between the hard and fast phases of 0, 180, 90, and 270 degrees. Any possible colour can be shown to have a particular phase of subcarrier. The colour saturation will be proportional to the amplitude of the subcarrier (Fig. 13).

As a bonus, if a part of the subject under observation owns a grey tint (as well as black or white, or any other intermediate grey tint) the modulated chrominance is momentarily and automatically cancelled while scanning that part. Furthermore, if there is no need to transmit in colour (like for example broadcasting an old B&W silent movie), the entire colour encoding circuitry (balanced modulators, subcarrier oscillator, etc.) could be turned off, transmitting so only the luminance component including synchronisms.

This technique is called *Quadrature Amplitude Modulation* (QAM).

This quite brilliantly solved the issue of joining the RGB components in a unique 'colour' signal. The remaining challenge was to find a method in which chrominance is added to luminance without causing any interference.

The first add-on was to introduce a trap to the chrominance signal for cutting-off the harmonic frequencies of the subcarrier. The next one was to choose suitable frequencies for the NTSC & PAL colour subcarriers also computed to reduce their visibility over the image and to minimize crosstalk between sync harmonics, audio subcarrier and the high-frequency details of monochrome images.

All the old NTSC sync frequencies for the incoming colour system had to be therefore recalculated in such way by using the following complex formulas:

$F^{H\text{-}NTSC} = (4.5 \times 10^6/286)$ Hz $= 15{,}734.27$ Hz,

new horizontal frequency (originally 15,750 Hz).

$F^{V\text{-}NTSC} = F^H/(525/2) = 59.94$ Hz,

new vertical frequency (originally 60 Hz).

$F^{SC\text{-}NTSC} = ((13 \times 7 \times 5)/2) \times F^H = (455/2) \times F^H = 3.579545$ MHz,
NTSC Subcarrier frequency.

Consequently the NTSC luminance bandwidth was low pass filtered to about 3MHz.

The new horizontal and vertical frequencies, slightly different from the NTSC monochrome standard, were within the tolerance ranges of sync revealers in the receiver and they were therefore acceptable.

For PAL the original F^H and F^V frequencies were maintained and therefore the following formula for calculating the PAL subcarrier frequency was applied:

$F^{SC\text{-}PAL} = (1135/4) F^{H\text{-}PAL} + (1/625) F^{H\text{-}PAL} = 4.43361875$ MHz,

where $F^{H\text{-}PAL}$ is 15,625 Hz.

Also in PAL the luminance bandwidth was reduced to about 4MHz.

Since the subcarrier is suppressed in chrominance signal and its phase should remain constant, in the receiver a very stable high-quality crystal oscillator running at colour system frequency (3.58 MHz for NTSC and 4.43 MHz for PAL) would be able to properly demodulate both chrominance components. However it was seen that this caused unstable hues, false colours and in worst cases random black and white images on colour televisions even frame-by-frame!

In order to counteract this huge problem, a sample known as *Colour Burst*, composed by 8-10 cycles of non-modulated subcarrier with a peak-to-peak amplitude equal to the height of the sync pulses, was introduced and sent at a regular rate inside each line. The place found for this sample was located immediately after each HS, i.e. inside the 5.8 μs back porches, the only available space which avoided problems on old B&W TV sets. This expedient furnished the decoder a reference signal to lock-in-phase with the local subcarrier oscillator and then the colour demodulator was able to correctly recover both chrominance components and the display device to show images with correct and stable hue.

Solved the boring phase-shifting trouble! But, unfortunately another phenomenon appeared during the NTSC development. Because of a non-linearity in the signal's path, i.e. a phase shift in either the transmission or inside the receiver itself and albeit the phase relationship between colour burst and chrominance generally remains constant, the relative phases of the aerial signal and the colour burst could undergo a defect called *Differential Phase Error*, in the form of a small yet stable delay or advance phase shift.

With very distant transmissions or weak RF signals this phase error could became very serious and therefore NTSC-only receivers have an external tint control (a user's control-knob called *Hue* on the TV set) that the viewer adjusts manually somewhere subjectively near skin's colour. Before digital migration doubtless this was (and still is...) for fifty years an uncomfortable operation for a NTSC user since the hue adjustment is often necessary during a TV zapping!

As we will see, this NTSC 'feature' is absent in PAL and SECAM televisions. In SECAM TV sets even the external colour saturation knob was eliminated!

In PAL the phase of V is reversed line after line, with corresponding reversal at the receiver. Because the scaled R-Y (V) is less attenuated than U, it is therefore less susceptible to a line-by-line switching processing. The resulting effect is a quasi-perfect line-by-line differential phase error cancellation; the correct hue will be reproduced at the receiver with the collateral effect of slightly reduced colour saturation. Furthermore the PAL decoder, via introduction in the encoding process a technique known as *PAL Switching* or *Swinging Burst* (consisting of alternating the phase of the colour burst by 180° line-by-line), will be able to distinguish the 'normal' and the 'reverse' V phase by means of the so-called *PAL Identification Circuit*.

'Simple' PAL decoders (called *PAL-S Decoders*) trust only on the averaging process to 'mislead' the eye. If these errors were serious, the result would be a visible artefact known as *Hannover Blinds* or *Hannover Bars*. 'Standard' PAL decoders (better known as *PAL-D Decoders*) use a 1-H (64μS) delay line to separate U from V. The result is that chrominance vertical resolution is reduced as a consequence of the line averaging process. More details are explained in the paragraph relative to the colour decoding process.

For a moment this alternating method was considered during the NTSC development but was rejected because at those times no cheap delay line was available. For this reason the PAL system was scoffed at by US engineers as 'Pay for Added Luxury'!

The modulated chrominance appears as a sine wave overlapped onto the monochrome signal when it is added with the sync, blanking and colour burst signals to form the so-called

Chapter 3. Colour Standard Systems

Composite Colour Video Signal or the abbreviation *CVBS*, the acronym for <u>C</u>olour, <u>V</u>ideo, <u>B</u>lank and <u>S</u>ync. This signal is finally ready to be elaborated, recorded, mixed, transmitted, etc. In Germany the German abbreviation *FBAS* (*<u>F</u>arb-<u>B</u>ild-<u>A</u>ustastung-<u>S</u>ynchron*) means the same. Its waveform is shown in Fig. 14.

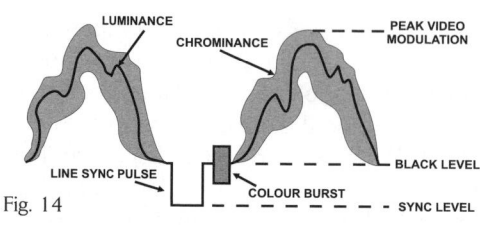

Fig. 14

At this point we can affirm that the amount of colour television information needed is, fortunately, much smaller than what would seem apparent and this enabled the development of a B&W compatible colour broadcast. In other words, Colour Television in its present form can only work due to the limitations of our human perceptual system!

We previously stated that luminance and chrominance had to be interleaved and, in order to easily separate the two video components at the receiver, the luminance bandwidth has to be reduced accordingly during the encoding phase to minimize interferences with chrominance subcarrier frequency, resulting in a reduction of image quality. In addition, since the QAM process introduced a small delay between luminance and chrominance, also a delay line of about 200 ns up to 800 ns was inserted in the luminance chain.

Despite these strategies, two artefacts, known as *Cross Colour* and *Cross Luminance*, could arise in the image due overall to imperfect Y/C separation filters employed in earlier or less expensive yet modern TV sets. *Cross Colour* creates a 'coloured *moirè* effect' overlaid on luminance repetitive patterns of an image; *Cross Luminance* generates 'herringbone patterns' especially evident around the edges of highly saturated colours. Improvements in electronics, especially from the digital sector, limited these issues to a quasi-invisible level.

In order to avoid any crosstalk between Y & C, especially evident in digitally-generated video signals, near 1980 some computer industries began to produce computers which generated a video signal made of two separate luminance and chrominance signals furnished on a single multi-pin socket named *Super Video Output*, shortly *S-Video Out*. The previous restrictive luminance bandwidth was increased significantly, improving image details correspondingly and eliminating cross colour and cross luminance defect. A display device equipped with an S-Video connector input was able to show such special bi-component video signal 'jumping' all the Y/C filters in order to directly drive the colour decoder stages.

The *Atari 800, Commodore 64* and *VIC 20* home computers were the first widely available devices to feature an S-Video output as well as standard composite video, displayed by proprietary TV monitor sets on which the picture quality was perceived to be appreciably enhanced especially in text mode display.

In 1987 JVC released its S-VHS video recording standard, which also introduced the 4-pin S-Video cable standard and the 4-pin mini-DIN connectors (or its IN/OUT professional version called *Y/C 7-pin Connector*) with a 75 Ω characteristic impedance.

However the S-Video format did not become as widely accepted as expected and remained in the niche standard high-end home theatre, semi and professional market. In the late 1990s some television equipment was manufactured and sold with S-Video ports with the consequent increasing of viable video devices embedded with such an output port as a better alternative to composite video output. In the early 2000s this format appeared also in some PC VGA graphic

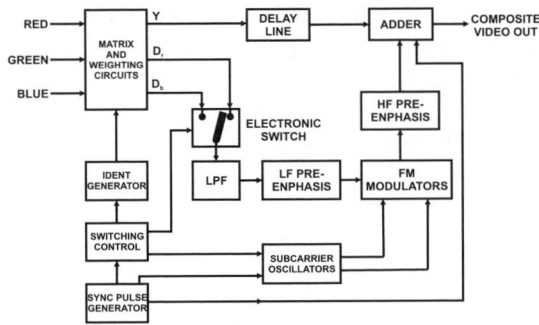

Fig. 15

cards as a low-cost video output for recording out-of-TV-standard VGA signals. Later, with the incoming diffusion of computerized editing video systems, also PC video capture cards were equipped with these ports both as inputs and as outputs.

3.4 SECAM System

The approach for encoding RGB colours in a unique video signal was completely different in the development of the SECAM system with respect to PAL or NTSC. The basic premise was that only one chrominance component is transmitted at any instant.

More exactly, the scaled versions of B-Y and R-Y, called in SECAM D_b and D_r respectively, modulate in frequency two subcarriers on alternate lines, almost eliminating in such way any interference as well as other benefits such as an insensitivity to phase distortion and amplitude frequency distortion.

SECAM matches CCIR625 requirements, like PAL, and was adopted by many countries; however, also many of them converted to PAL due to the abundance of professional and consumer PAL equipment and for several severe technical problems for its implementation, as we shall see (Fig. 15).

Like the other two colour systems, for backward compatibility, the luminance signal Y is transmitted in full and is derived from RGB signal in proportions equal to NTSC & PAL:

Y = 30% Red + 59% Green + 11% Blue.

The colour differences signals D_b and D_r were instead scaled in this way:

$D_b = 1.505(B-Y)$;

$D_r = -1.902(R-Y)$,

where the minus sign means inversion of polarity.

The idea of the inventor, *Henri de France*, was based on the observation that the human eye is quite insensitive to colours with respect to grey shades, so that only half the vertical resolution could be sufficient for transmitting colours.

More precisely, instead of transmitting both colour difference signals at the same time, as in NTSC and PAL, in SECAM one line carries D_r signal and the next one the D_b signal. In order to recover all the informations needed to make a full colour picture, you should delay one line by means of a 1-H (64µS) line memory store (from which SECAM derives its name, *Séquentiel Couleur À Mémoire*, i.e. 'sequential colour in memory') to have contemporaneously luminance and colour informations.

After line-by-line switching, the D_b and D_r signals undergo two processes before modulating in frequency on two different subcarriers. The first one is band limiting to approximately 1.5 MHz for both signals and the second one a pre-emphasis after which D_b and D_r are amplitude limited and clamped to establish a black level reference.

No crystal quartz subcarrier oscillator is employed in SECAM. The frequencies of subcarriers are PLL-locked with HS of frequency F^H (15,625 Hz) in this way:

$F^{Db} = 272 \ F^H = 4.250000$ MHz,

driven by D_b signal;

$F^{Dr} = 282 \ F^H = 4.406250$ MHz,

driven by D_r signal.

Both subcarriers must own ±2 kHz of tolerance and nominal modulation shifts are ±280 kHz and ±230 kHz for D_r and D_b respectively. In addition the phase of the subcarriers is also reversed by 180 degree on every third line and between each field to further reduce subcarrier visibility. Note that phase does not carry any colour information.

Unlike PAL and NTSC, in SECAM both subcarriers are always active also in grey parts of an image revealing their presences under low luminance or weak RF reception with an artefact known as *SECAM fire* or *flame*.

Some expedients were studied and developed to attenuate this defect, such as a subcarrier pre-emphasis by changing the width of the subcarrier as a function of frequency deviation, but the problem remained and still remains.

Like PAL, SECAM needs an identifying pulse for decoding the exacting line-switching sequence. The method employed since 1979 was a 'burst' after each HS pulse which gives a particular polarity line-by-line for properly switching a synchronization circuit in the decoder.

Of course, SECAM CVBS is generated adding subcarrier data to the luminance alongside composite syncs, blanking and burst signals.

SECAM has several variants, depending on the video bandwidth and placement of the audio subcarrier. The video signal has a bandwidth of 5.0 or 6.0 MHz, depending on the specific SECAM version standard.

3.5 The relative merits and demerits of TV standards

The differences between each of the main TV systems are not as clear as one might imagine at first sight. While NTSC has a reputation of poor colour accuracy, this is only true of television and a video format that has some advantages over other systems. Recall that all of these systems are actually compromises and many efforts have been made over the years to address deficiencies in each system.

Below I will try to create as objectively as possible a comparison offering advantages and disadvantages of each analogue colour system.

3.5 The relative merits and demerits of TV standards

3.5.1 NTSC/525-lines pros

* *Higher frame rate* - Use of 29.97 frames per second reduces visible flicker.

* *Precision colour editing* - With NTSC you can edit at any 4 field boundary points without disturbing the colour signal.

* *Less inherent picture noise* - Almost all pieces of video equipment achieve better signal to noise characteristics in their NTSC/525-lines form than in their PAL/625-lines.

3.5.2 NTSC/525-lines cons

* *Lower number of scan lines* - Reduced clarity on large screen TV sets; the scan lines became more visible.

* *Smaller Luminance Signal Bandwidth* - Due to the use of the colour sub-carrier at 3.58MHz, image defects such as *moiré*, cross-colour, and point interference become more pronounced because of the increased risk of interaction with the black and white picture signal at the lower sub-carrier frequency.

* *Susceptibility to Hue Fluctuation* - Changes in the colour subcarrier phase cause changes in the displayed colour, requiring that the TV receiver is equipped with a hue adjustment to compensate. That is why some people jokingly refer to NTSC as an acronym for 'Never Twice the Same Colour' or simply 'Never The Same Colour'.

* *Undesirable Automatic Features* - Many NTSC TV receivers feature an *Auto-Tint* circuit to make hue fluctuations less visible to uncritical viewers. This circuit will change all colours to approximate the 'skin' colour in a tone of 'standard' flesh tone, and then hide the effects of hue fluctuation. This means, however, that a range of colour shades cannot be correctly displayed. On recent televisions (excluding cheapest) there is often found a 'turn-off auto-tint' function button.

* *Fully incompatible with SECAM & PAL.*

3.5.3 PAL/625-lines pros

* *Greater Number of Scan Lines* - More picture details.

* *Wider Luminance Signal Bandwidth* - The placing of the colour sub-carrier at 4.43MHz allows a larger bandwidth of monochrome information to be reproduced than with NTSC.

* *Stable Hues* - Due to the reversal of sub-carrier phase on alternate lines, any phase error will be corrected by an equal and opposite error on the next line, correcting the original error. In early PAL implementations it was left to the low resolution of the human eye's colour abilities to provide the averaging effect; afterwards it was done with a delay line.

* *B&W visual compatibility with SECAM TV sets.*

3.5.4 PAL/625-lines cons

* *More Flicker* - Due to the lower frame rate, flicker is more appreciable on PAL transmissions, particularly for people used to viewing NTSC footages.

* *Lower Signal to Noise Ratio* - The higher bandwidth requirements cause PAL equipment to have slightly worse signal to noise performance than equivalent in NTSC versions.

* *Loss of Colour Editing Accuracy* - Due to the alternation of the phase of the colour signal, the phase and the colour signal only reach a common point once every eight fields (or four frames). This means that edits can only be performed to an accuracy of ±4 frames (8 fields).

* *Variable Colour Saturation* - PAL colour, which cancels the differences between two successive line signals, can reduce the colour saturation and hue stability. Fortunately, the human eye is much less sensitive to variations in saturation than for hue variations. The PAL system also has a nickname. Because it was adopted by a large majority of world countries, it came to be called 'Peace At Last', and numerous variants like 'Pray And Learn', 'Pray And Look', 'Perfection At Last', 'People Are Lavender', 'Picture Always Lousy' or 'Pay Another License'!

3.5.5 SECAM/625-lines pros

* *Stable Hues and Constant Saturation* - SECAM shares with PAL the ability to make images with the right tone, and makes a step forward in ensuring consistency of colour saturation as well.

* *Higher Number of Scan Lines* - SECAM shares with PAL/625-lines, higher than NTSC/525-lines.

* *No crystal quartz colour carrier oscillator is necessary in the receiver.*

* *B&W visual compatibility with PAL TV sets.*

3.5.6 SECAM/625-lines cons

* *Greater Flicker like PAL.*

* *Mixing of two synchronous SECAM colour signals is not possible* - When a composite signal involving luminance and chrominance is faded out in a studio operation it is the luminance signal that is readily attenuated and not the chrominance. This makes the colour more saturated during fade to black. Thus a pink colour will change to red during fade-out. For this reason most TV studios in SECAM countries work in PAL and transcode in SECAM before broadcasting. Even if SECAM permits more simplified recording circuits on magnetic tape recorders, more advanced home systems such as SuperVHS, Hi-8, and Laser Disc work internally in PAL and transcode on replay in SECAM market models.

* *Patterning Effects* - The FM subcarriers cause patterning effects even on non-coloured part of objects.

3.5 The relative merits and demerits of TV standards

* *Lower Monochrome Bandwidth* - Having one of the two colour sub-carriers at 4.25MHz (in the French version), a lower bandwidth of monochrome signal can be employed.
* *Incompatibility between different versions of SECAM* - SECAM being at least partially politically inspired, has a wide range of variants, many of which are incompatible with each other. For example between French SECAM which uses FM subcarriers, and MESECAM which uses AM subcarriers.
* *Smearing of intense saturated colours* (called SECAM 'flame' or 'fire') under weak RF reception conditions.

Like its competitors, the French system could not escape nicknames. Developed in the time of General de Gaulle and his refusal to unconditionally accept the American leadership in NATO, the SECAM became jokingly the acronym for '*Something Essentially Contrary to the American System*' or sometime also for '*Silly Europeans Continuing American Mistakes*' or even '*Supreme Effort Contre les AMericains*'!

In conclusion, it can be stated that all TV systems are compromises. In fact individually each system owns specific peculiarities which outclass the other two systems and it is difficult to justify the absolute superiority of one system compared to others, also taking also into account that their adoption were determined by political and economic decisions. We can therefore safely conclude that the three colour systems could coexist.

TAB.1 - TV Picture Format/Colour System Combinations

The following table describes some characteristics of the standard analogue video formats in common use:

Mode	Signal Name	Frame Rate (fps)	Vertical Line Resolution	Line Rate (Hz)	Colour Subcarrier (MHz)
Mono	RS-170	30	525	15,750	
Colour	NTSC	29.97	525	15,734	3.58
Mono	EMI CCIR	25	405 625	10,125 15,625	
Colour	PAL	25	625	15,625	4.43
Mono	---	25	819	20,475	
Colour	SECAM	25	625	15,625	4.25/4.4

Chapter 3. Colour Standard Systems

3.6 Other analogue colour television systems

After the Italian ISA colour system, patented in 1975 but never produced in volume, other colour systems were developed, like Russian *SECAM-IV* (called also *Linear NIR*), *Extended PAL*, British *Clean PAL, PAL+*, and minor PAL or NTSC sub-standards, adopted across the world for terrestrial transmission alongside the legacy systems.

During the eighties, some European television satellite companies attempted to define a common standard for their upcoming broadcasts, with the intention to eliminate most drawbacks of composite video systems and at the same time to improve picture quality as well as to introduce digital sound.

MAC (*Multiplexed Analogue Components*) system, followed by its numerous sub-systems, was the first television standard to break with 'old' traditional standards by temporarily entering the digital domain.

D2-MAC is the most well-known spin-off of these hybrid systems, which unfortunately did not reach a great success primarily due to its late adoption for a large public acceptance before the anticipated development of MPEG-based digital broadcasts.

Since each individual viewer had their own satellite receiver, MAC was no more constrained to a complex colour encoding or monochrome backward compatibility thus adopting a new technique known as *Time Division Multiplexing* (*TDM*).

However, because the satellite transmitted pictures must be reconstructed on a standard CCIR625 display device, usually connecting a SCART RGB cable between satellite receiver and TV set, MAC retains the line scanning parameters of 625 lines/25 fps/Interlace 2:1 so that it could be encoded to PAL or SECAM without a standards converter. No NTSC (RS170) equivalent satellite system was designed or developed.

On the transmitter side a TV line is previously split in its two main components, luminance and chrominance whose U and V components are transmitted alternately from line to line in order to reduce the necessary bandwidth, similarly to SECAM.

Next, both components Y & C are time-compressed by a factor of $\frac{2}{3}$ for Y and $\frac{1}{3}$ for C, scrambled if required, digitized at an elevated rate, stored in a memory and finally included as consecutive and separated packets over one line duration.

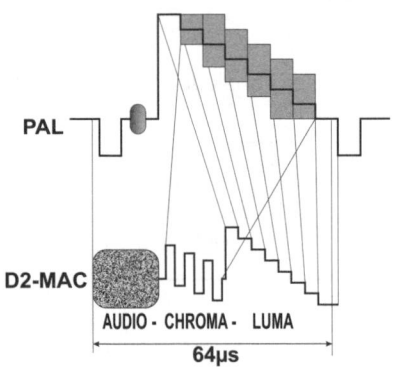

Fig. 16

The space between the two Y/C packets, where normally we relieve sync and blanking pulses, is occupied by a 'duo-binary data' burst (hence the 'D2' in D2-MAC) which carries all the informations regarding the digital sound, synchronization, teletext, captioning, picture format ratio (4:3 or 16:9), and in addition, for pay TV programs, the bits for decrypting scrambled transmissions (Fig. 16).

The MAC system's numerous variants were employed exclusively for satellite transmission. With the exception of dedicated satellite receivers which converted aerial frequencies in separate component outputs to directly drive a PAL TV set, no other video equipment was designed for MAC. As a result, before MPEG-based satellite and terrestrial digital transmission, all these systems had nothing more than a shadowy existence.

3.7 Colour Encoders

Now it is the turn of explaining how to generate electronically composite video signal (CVBS) starting from three RGB component signals. Historically NTSC was the first fully electronic colour system developed during the early fifties years, when electronics consisted of electronic tubes assembled by hand on a metallic chassis to create working circuits. The invention of transistor and integrated circuits simplified the design, construction and assembly of printed circuit boards both for colour encoders and decoders, with consequent miniaturization of such equipment now to be based on modulator ICs like the *TBA520, MC1496, MC1596, TCA240*, etc. The low-end, cost-driven consumer market concentrated its efforts to produce encoders and decoders on a single monolithic chip. For high-end video equipments, such as broadcast and professional cameras, the encoders were always made up of discrete components having many fine-tuning sites or calibrations which necessitated professional, dedicated video instruments, such as vectorscopes and waveform monitors.

3.7.1 Digital inputs encoder chips

Mullard (and its partnership Philips) was one of the first companies to introduce an one-chip colour encoder, the *TEA1002* (18 pin DIL packaged), described as a PAL colour encoder and video adder chip with an internal 8.86 MHz oscillator thus generating 4.43 MHz R-Y and B-Y waveforms. For TV games systems, a 3.54 MHz clock output is provided which could feed a microprocessor. The TEA1002 accepts timing signals (composite syncs, burst gate, PAL switch and composite blanking) from an external SPG and 4-bit binary coded logic inputs give colour information from logic combinations at TTL levels. The resulting output, which has an adjustable DC level, is a 16 colour (including black and white) composite video signal, based on 75% chrominance. Alternatively, with INV input connected to ground and the DC adjustment disabled, the TEA1002 can be used as a general purpose video encoder providing standard 95% chrominance from RGB logic inputs, suitable for applications such as add-on teletext. *E.L.C. East London Components* (www.elclondon.co.uk) has listed this component on its website.

The *TEA2000* (18 pin DIP), from Philips, was an evolution of the TEA1002 (Fig. 17). It furnished 2-bits for each colour input giving in total $2^{(2+3)}=64$ colours including white, black and 2 grey tones. It only requires syncs (negative polarity) and a set of TTL level RGB signals to drive it. BURST GATE and PAL switch signals are generated internally starting

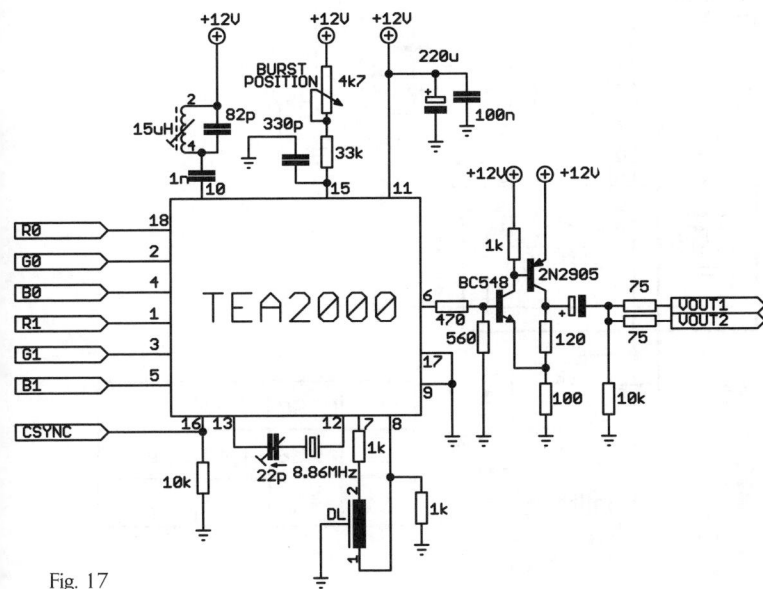

Fig. 17

Chapter 3. Colour Standard Systems

from composite syncs. As the TEA1002, a suitable built-in crystal quartz oscillator (7.16 MHz for NTSC or 8.86 MHz for PAL) and a divider by 2 furnish the correct subcarrier frequency. The other requirement necessary for correct PAL operation is to leave pin 14 open-circuit otherwise grounded for NTSC. The blanking input (pin 17) must be kept high during sync and colour burst, unless all colour inputs are low at this time. In Fig. 17 there is a typical application.

National Semiconductors is another manufacturer that introduced the 'TV Video Modulator' *LM1889* (18 pin DIP package), which allowed video information from VTR, videogames, test equipment or similar sources, to be displayed on black and white or colour TV receivers.

It consisted of a sound subcarrier oscillator, chroma subcarrier oscillator, quadrature chroma modulators, and RF oscillators and modulators for two low-VHF channels and therefore its fundamental use was intended for non-broadcast applications. Via three inputs R-Y, B-Y & Y, and through a subcarrier crystal quartz which ensures the correct generation of chroma signal, the LM1889 is capable of generating composite video at 1Vpp amplitude. The LM1889 was very popular in Europe because it gave a compatible PAL video output to 8-bit computers like the *Sinclair ZX Spectrum* and *DAI Personal Computer*.

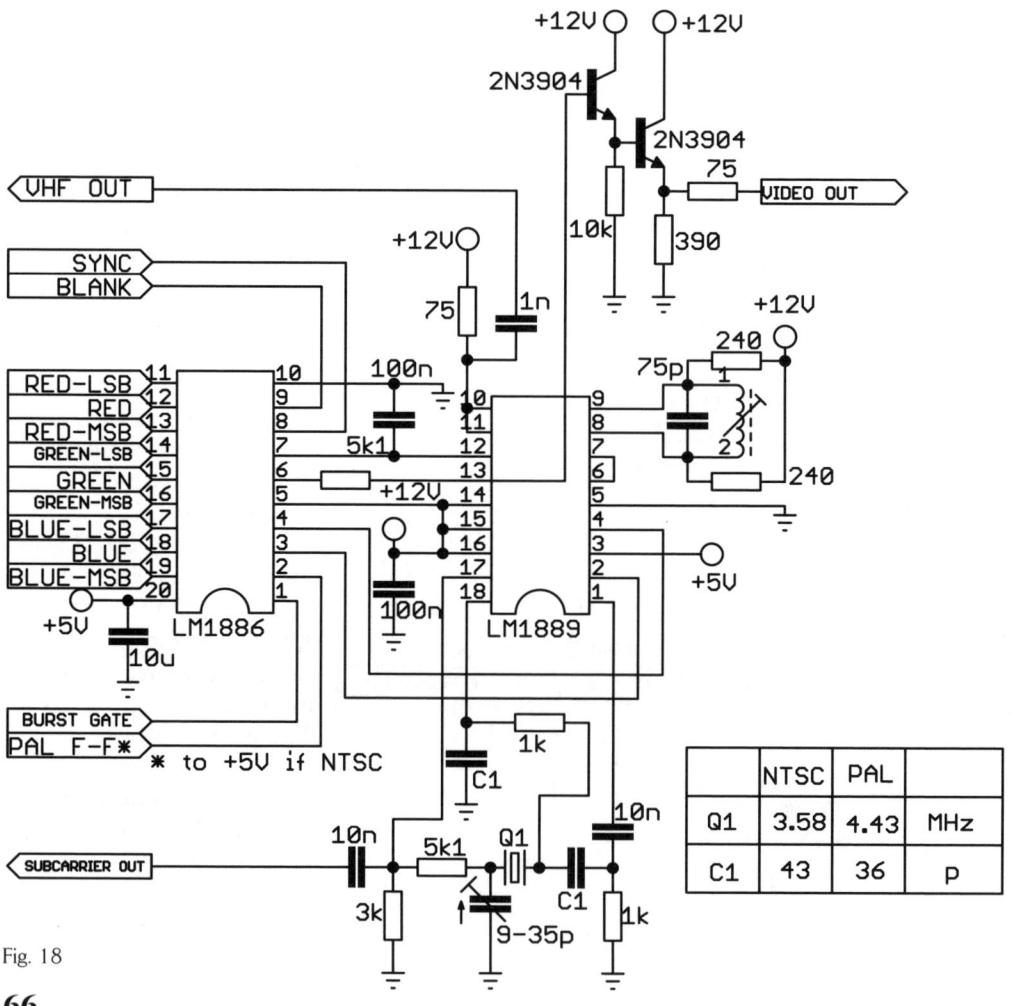

Fig. 18

3.7 Colour Encoders

National produced a LM1889 companion, the *LM1886* (20 pin DIP), described as a *"TV Video Matrix D to A converter which encodes luminance and colour difference signals from 3-bit red, green and blue inputs"*. Note that because the LM1886's RGB inputs are digital, these are not applicable to analogue video sources such as those, for example, generated by video camera devices. The 3-bit RGB inputs are however able to create $2^{(3+3)}=512$ colour combinations by setting them individually to '0' (ground) or '1' (+5V) logic level. After internal matrixing, the LM1886 delivers Y, B-Y & R-Y and the DC reference all at the correct levels to interface with the LM1889 so to produce a standard PAL composite video signal. In Fig. 18 there is a typical application for both National chips.

If you want to connect SPG625 to easily drive these chips so to create a CVBS PAL signal, tie CBLK, CSYNC and PAL signals to the correspondent LM1886's inputs BLANK, SYNC and PAL F-F, or to TEA1002's inputs CBLNK (previously inverted by a logic gate), CSYNC and PAL switch, or to TEA2000's inputs CBLNK (previously inverted by a logic gate too) and CSYNC. The necessary extra BURST GATE signal for correctly operating both LM1886 and TEA1002 can be generated starting from pin 9/IC8 of SPG625 (which offers the horizontal sync HS of positive polarity) by means of a dual monostable IC type 4538. The first monostable creates a delay of 0.6 µS after HS falling edge and the second one a pulse of 2.25 µS (equal to 10 subcarrier periods), which represents the real burst gate pulse, present at pin 10 (output Q, positive polarity for TEA1002) and 9 (Q\, negative polarity for LM1886) of IC1B (see Fig. 19). The LM1889 is still available from some international dealers such as E.L.C. (www.elclondon.co.uk) and TME - Transfer Multisort Elektronik (www.tme.eu).

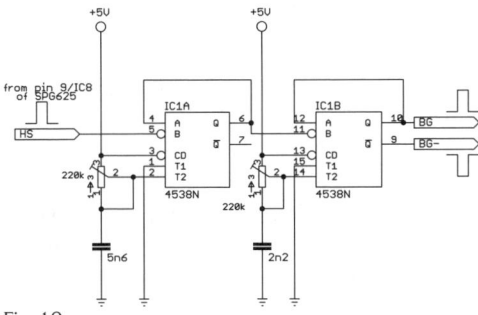

Fig. 19

Sanyo produced the *LC78011E* (packaged as QIP48E), a CMOS IC that integrates a digital RGB encoder and a synchronizing signal generation circuit on a single chip. A video CD or CD-G system can be formed using the LC78011E together with an MPEG video decoder or a CD-G decoder, respectively. The LC78011E supports the four input formats: R, G, and B (8 bits each); Y, U, and V (8 bits each); Y and UV (8 bits each; the UV input is a multiplexed input); R, G, and B (4 bits each in CD-G input support mode) and a 4-bit OSD (*On Screen Display*) input support. Starting from external synchronizing signal inputs HSYNC, CSYNC, BLANK and a $4F_{sc}$ system clock (NTSC mode: 14.31818 MHz, PAL mode: 17.734475 MHz), it can create the following two video signal output formats: luminance signal (Y) and chroma signal (C) outputs or composite video signal (CVBS) output in PAL or NTSC formats. In Fig. 20 there is the system block diagram. Other information can be found in its datasheet freely downloadable from www.datasheetcatalog.com.

A similar device is the *BU1424K* from the Japanese ROHM, described as an IC that converts digital RGB/YUV input to composite (NTSC & PAL), luminance and chrominance signals, and outputs the results for applications as video interfaces for VIDEO-CDs and CD-G decoders.

Analog Devices produces some digital encoders designed to cover the range from handsets to high-end video equipment. The *ADV717x* and *ADV73xx* families of digital video encoders output multiple analogue formats conforming to international SDTV & HDTV video standards.

Chapter 3. Colour Standard Systems

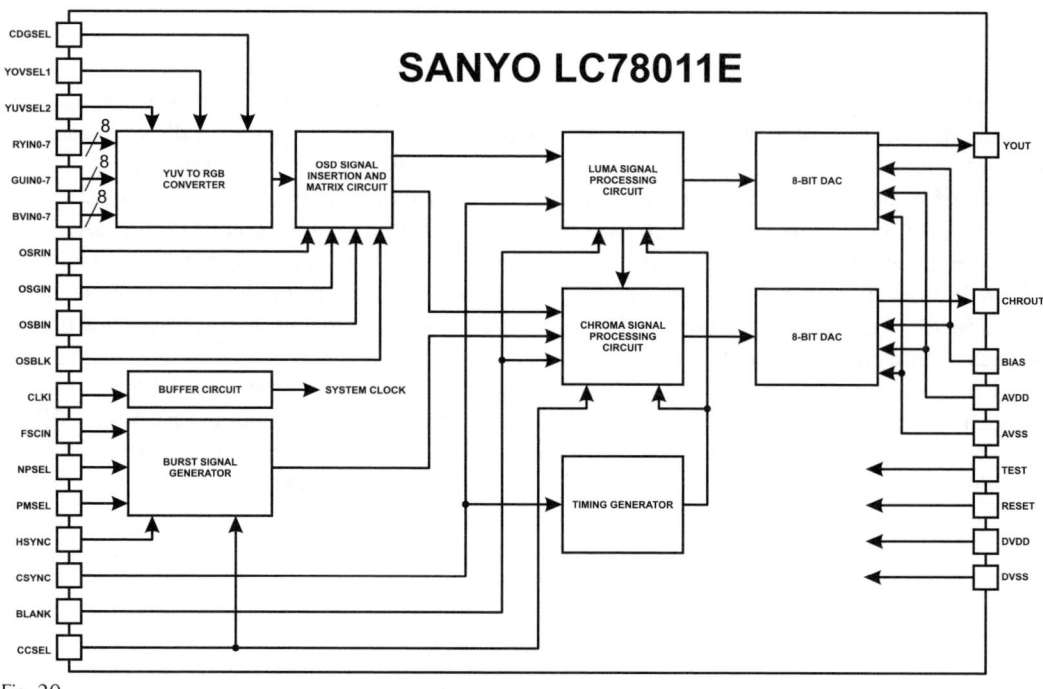

Fig. 20

8 to 14-bit DAC options offer designers flexibility and performance, meeting the most demanding product requirements. More information is available on Analog Devices website (www.analog.com). Some of them are available from TME (www.tme.eu).

3.7.2 Analogue input encoder chips

One of the most revolutionary colour encoder chips was the Motorola *MC1377P* (20 pin DIL; a SMD version was also available) that claimed to operate to standards that permitted quality TV camera applications, computers and TV games consoles. This device was somewhat unique for its day because it accepted red, green and blue analogue signals and composite synchronisms, and encoded them into a composite video signal CVBS in either PAL or NTSC formats. The IC contains an on-board subcarrier reference Colpitts oscillator that may optionally be driven from an external oscillator, a voltage controlled 90-degree phase shifter, two double sideband modulators and blanking level clamps. It was employed in the *Sinclair ZX128 & QL*, *Commodore Amiga 1000* and the colour external modulator *A520* for Amiga computers, and in many low cost video equipments. In Fig. 21 there is a typical application.

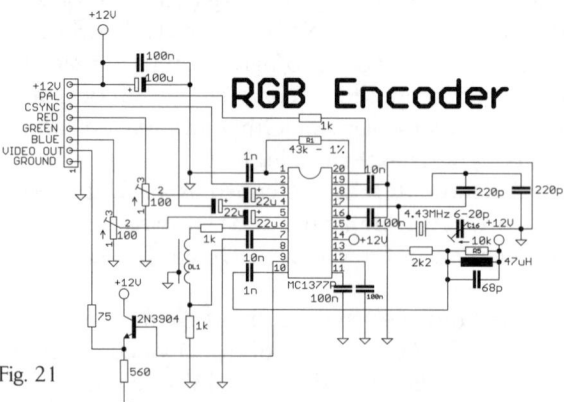

Fig. 21

R-Y, B-Y and -Y signals generated in the input matrices are DC clamped to black level from a sync clamping procedure. Burst generation is provided by a charging capacitor ramp on pin 1, but only a small part of it is used with the result that sufficient accuracy is achieved with the use of passive components. Burst size is set internally to correspond to the level of syncs, allowing for a loss of 3 dB bandwidth of the chrominance. The colour reference burst is generated internally by gated mixed syncs. An interesting note is that composite video can be used directly as a sync signal, if it meets the criteria for entry of synchronization. More tips about how to perform this last operation are explained on its datasheet freely downloadable from *Freescale* (www.freescale.com), a manufacturer created by the divestiture of the *Semiconductor Products Sector of Motorola*.

The chroma signals saturate at 1.0 Vpp so a bandpass filter (from pin 13 to pin 10) can be accomplished with a standard bandpass coil arrangement. The 400 ns luminance delay line DL1 is of the same standard as that found in a colour TV performing yet the opposite function in the encoder. Experience told me that a 400 ns delay was too much and actually causes a small colour registration error on the screen; a midway value (if available) can be acceptable in most cases or the delay line can be eliminated completely in worst cases.

The RGB inputs are AC coupled to simplify interfacing with a variety of sources, and 1 Vpp signals give full colour saturation. However in practice it was found that, although it is true for the Red and Blue inputs, Green required only about 0.8 V, so at least two trimmers on Red & Blue inputs have to be use for balancing out in order to get a 'clean monochrome colour', i.e. without chrominance information. The composite blanking must be included in the RGB information signals otherwise, if these are present during blanking, correct functioning of MC1377P can be affected.

Pin 20 is used to switch between standards: left open for PAL, for performing the reverse line chroma-swinging signal generated internally by gated mixed input sync, or must be grounded for NTSC. It can be driven externally by a $F^H/2$ signal for locking the right PAL alternating phase by an external $F^H/2$ source.

The MC1377 has a composite output level of 2 Vpp which is elevated above ground. In order to provide a standard 1 Vpp signal over 75 Ω, an emitter follower stage is provided which also serves as a buffer to protect the IC from unwanted signals on its output pin.

The circuit generally requires no alignment to work; however small adjustments may be necessary. After 5/6 minute warm-up time, connect a frequency counter to pin 17 and adjust C16 for 4.433619 MHz for PAL or 3.579545 MHz for NTSC.

Using an oscilloscope the burst position can be adjusted for small changes in the value of R1, or similarly the chroma level can be still further trimmed by altering R5.

If a colour hue or tint is present (especially evident in white and grey areas), the level of R or B inputs must be calibrated to reduce it slightly until the hue has disappeared. The best way is to join the three RGB inputs to be fed from a common signal of 1 Vpp and calibrate the two trimmers so to minimize the residual chrominance.

Calibration is undemanding for all but the most exacting users. For the latter key points should be mentioned:

* Subcarrier suppression is poor and it is very noticeable on colour bars. An attempt was made to reduce this by phase adjustment at MC1377's pin 19 following its datasheet guidelines, but no real improvement could be effected.

* There is some residual colour patterning in greyscale parts of the picture.

* A small chroma-luma registration error is present caused by having to use a 400nS delay line. I prefer to recommend a 200nS or less.

* A ghost vertical line is visible on the left side of a black raster. Normally it is unimportant

Chapter 3. Colour Standard Systems

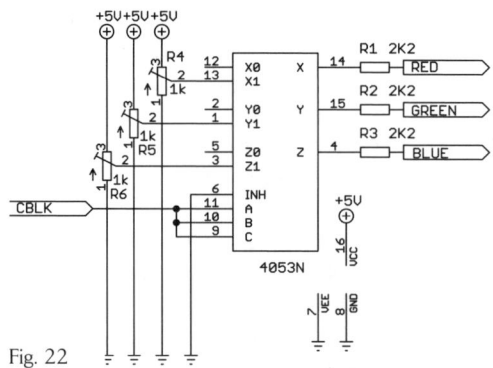

Fig. 22

but under some conditions it is noticeable especially when increasing the brightness of the TV set. It is caused by the ramp internal circuitry and there is nothing to do to eliminate it (it is a design flaw!).

To connect SPG625 to MC1377P it is sufficient to directly interconnect the two CSYNC lines which create a PAL compliant video composite black raster: this is commonly called *Black & Burst*. For colours, an additional circuit has to be used, visible in Fig. 22. The CBLK signal coming from SPG625 will drive the A, B & C control inputs of the triple 2-Channel Analogue Multiplexer/Demultiplexer IC1 (4053). Three trimmers (R4, R5 & R6) connected each one between +5V and ground, feed the corresponding R, G & B inputs of MC1377P crossing IC1 via X1, Y1 & Z1 inputs, thus creating three individual signals simultaneously interrupted by CBLK to satisfy MC1377's correct encoding processing. By rotating the trimmers between positions, practically any colour can be reproduced. A fussy experimenter can substitute R1, R2 & R3 with three trimmers of the same value for fine adjusting. In particular R2 has to be calibrated to balance RGB levels thus to eliminate any residual subcarrier signal on a white raster, by turning the R & B trimmer cursors towards +5V to get 1Vpp at the corresponding MC1377's R & B inputs. An oscilloscope can be a valuable aid for this.

Having said all that, this little circuit offers a very simple yet practical encoder which may be fitted almost anywhere or combined on a single printed circuit board together with a single-chip master sync pulse generator as those ones listed in the previous chapter. Based on a typical power supply of 12VDC the circuit's consumption is less than 100mA under a 75 Ω load.

The MC1377P had an enormous success due to its relatively low cost, simple design and great reliability. However Motorola successively produced two more professional encoder chips, the *MC1378* and *MC13077*.

The former was extensively employed in professional boards such as GVP's *Impact Vision 24* (a superimposer and frame grabber card for high-end Amiga computers), because it collected some unique features expressively dedicated to video-specialized microcomputer add-ons. In fact the MC1378 was the first bipolar composite video overlay encoder and microcomputer synchronizer. The MC1378 contains the complete encoder function of the MC1377, i.e., quadrature colour modulators, RGB matrix, and blanking level clamps, plus a complete complement of synchronizers to lock a microcomputer-based video source to any remote video source. The MC1378 can be used as local system timing and encoding source, but it is most valuable when used to lock the microcomputer source to a remotely originated video signal. It contains all needed reference oscillators as well as capability of PAL or NTSC operating mode, 625 or 525 line standards. Wideband full-fidelity colour encoding was possible using minimal external components. It was designed to operate from a 5V DC supply in local or remote modes of operation also with non-standard video signals. The MC1378 was distributed in 40 pin DIL or SMD PLCC44 plastic package. In Fig. 23 there are shown the two MC1378's operation modes.

In the same MC1377/MC1378 philosophy, Motorola designed the *MC13077*, a furthermore evolution of the previous chips.

3.7 Colour Encoders

MC1378

Remote Mode

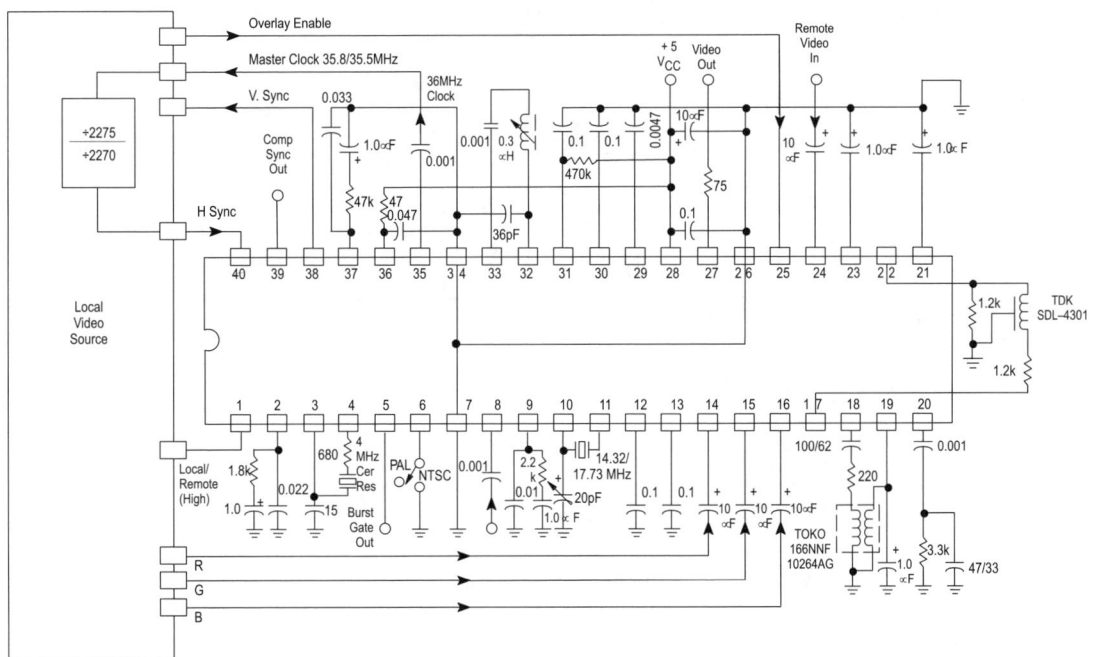

Local Mode

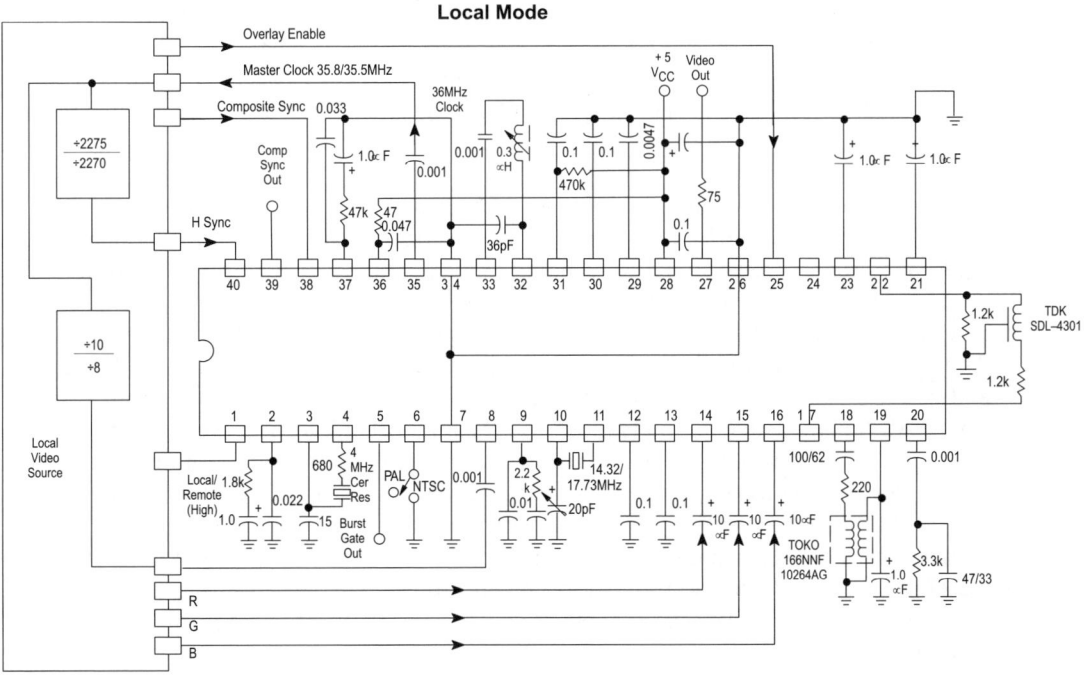

Fig. 23

Chapter 3. Colour Standard Systems

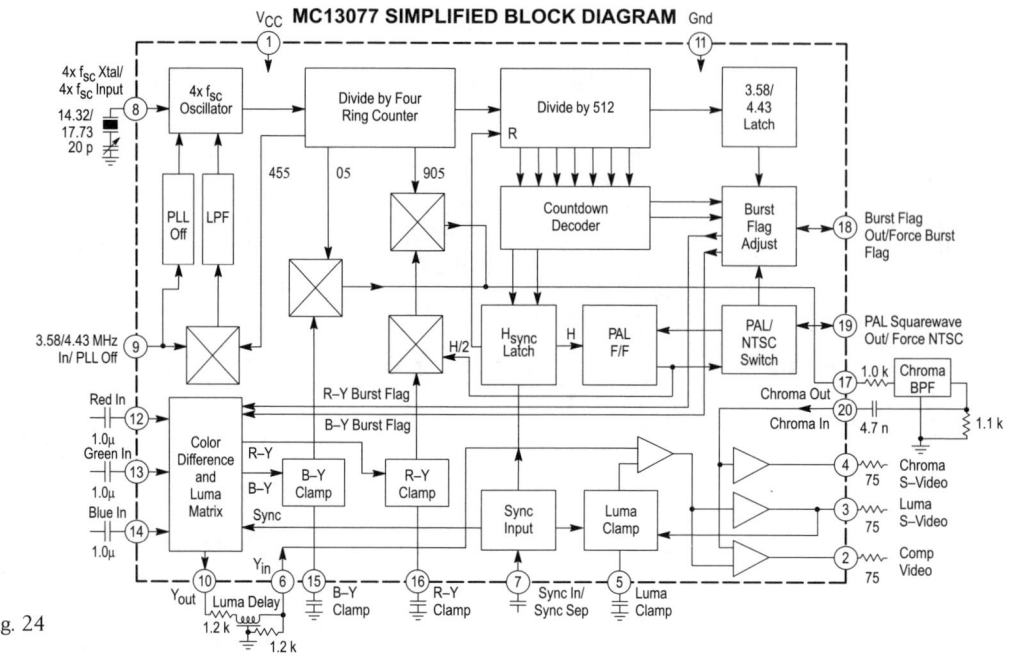

Fig. 24

The MC13077 was described from Motorola as an advanced high-quality RGB/YUV to NTSC/PAL encoder with composite video and S-Video outputs. The IC integrates the colour difference and luma matrix circuitry, chroma modulators, subcarrier oscillator, and logic circuitry to encode component video into a composite video signal compatible with the NTSC/PAL standards. The high degree of integration saves board space and cost, as only passive external components are required for operation.

After the requirements of many adjustments needed on early discrete encoders, it is truly remarkable (and bemusing...) to see the quality of the signal that emerges from this chip in terms of subcarrier leakage, precise matrixing, perfect quadrature modulation, and all that without any external calibration!

The simplified block diagram visible in Fig. 24 shows the use of a $4F_{sc}$ oscillator (14.32 MHz for NTSC, 17.73 MHz for PAL) as an internal clock to generate accurate subcarrier phase pulses and internal pulses for free-running operation. However, there is a built-in Phase Locked Loop (PLL) circuit allowing $4F_{sc}$ to be locked to an external subcarrier applied to pin 9. Similarly, the internal Burst Gate and PAL switch pulses can be superposed or reset externally.

The sync input (pin 7) includes a sync separator too and so the reference could be a video signal if convenient. For PAL encoding, serrated vertical sync pulses are necessary to maintain the right PAL Square Wave sequence, but for NTSC a 'clean' vertical sync is acceptable. In NTSC mode, the PS output (pin 19) is grounded; in the PAL mode the internally generated PAL pulse can be reset to phase lock to an external PAL signal by momentarily grounding the pin 19. For that the small circuit in Fig. 25 can be helpful.

This IC operates off a standard +5.0 V supply and typically requires less than 75 mA, making it useful in experimental and professional environments. The three outputs CVBS, Y and C outputs are each emitter followers capable of driving video into 75Ω loads in series with a 75 Ω external resistor.

The data sheet recommends a chroma bandpass filter type TOKO H286BAIS–6276DAD

3.7 Colour Encoders

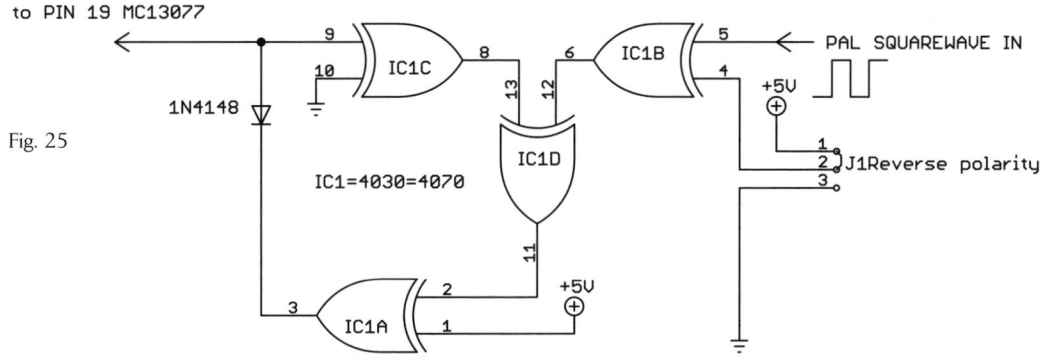

Fig. 25

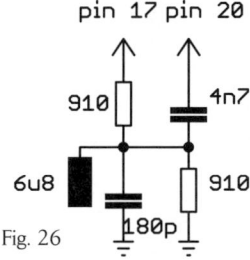

Fig. 26

for NTSC or TOKO H286BAIS–4963DAD for PAL. Because the TOKO filters are very hard to find, a simple network filter can be employed instead. In Fig. 26 there is a chroma filter designed for PAL only. By modifying values discussed it can be adapted for NTSC.

The delay of this filter leads to the requirement of a 400nS delay in the luminance path. Just like for the MC1377, that delay time seems to be too much for a good reproduction in practice, so all the considerations previously relevant for the MC1377 are valid also for the MC13077.

Although the MC13077 is specified for RGB inputs, YUV can be accepted by feeding Y into the RGB inputs, U and V directly into the chroma modulators using the pins that normally connect the clamp capacitor on pin 15 (U input) and 16 (V input). For more details on this mode of operation, see the data sheets available on the Freescale website. Note that the output level of synchronisms is not adjustable and is related at the NTSC standard of 280 mV, rather than the PAL standard of 300 mV. Normally this small difference level is irrelevant.

The MC1378 and MC13077 were supposed to replace the old MC1377 but they have not had the same success.

The MC1377P was produced until 2000 but it is still available from *Futurlec* (www.futurlec.com) and E.L.C. (www.elclondon.co.uk) as well as TME (www.tme.eu); the *NTE879* from NTE Electronics is a perfect MC1377 replacement normally in production. The MC1378 and MC13077 are now hard to find, albeit some specialized dealers in discontinued products can get them.

Sony also marketed some PAL/NTSC colour encoder chips among which the *CXA1145*, *CXA1645*, *CXA2075* and *V7040* are notable. The first three chips are quite equivalent and Sony had licensed them to other manufactures including ROHM (marked *BH7236*), Samsung (*KA2195D*, *KA2198BD*), Fujitsu Semiconductor (*MB3516A*), Mitsumi (*MM1268*) and others. The CXA1645P is still available from TME (www.tme.eu) but it is quite expensive.

The *CXA2075M* is an advanced version of the CXA1645 encoder IC which converts analogue RGB signals to a composite video signal. This IC has various pulse generators necessary for encoding. Composite video and Y/C outputs are obtained just by inputting composite sync, subcarrier and analogue RGB signals. It is best suited to image processing of personal computers and video games.

The *CXA1145* owns a built-in subcarrier oscillator for which it was primarily selected by Commodore to give a high-quality PAL CVBS output to its low-end *Amiga 1200* computers. In Fig. 27 there is an application note.

Chapter 3. Colour Standard Systems

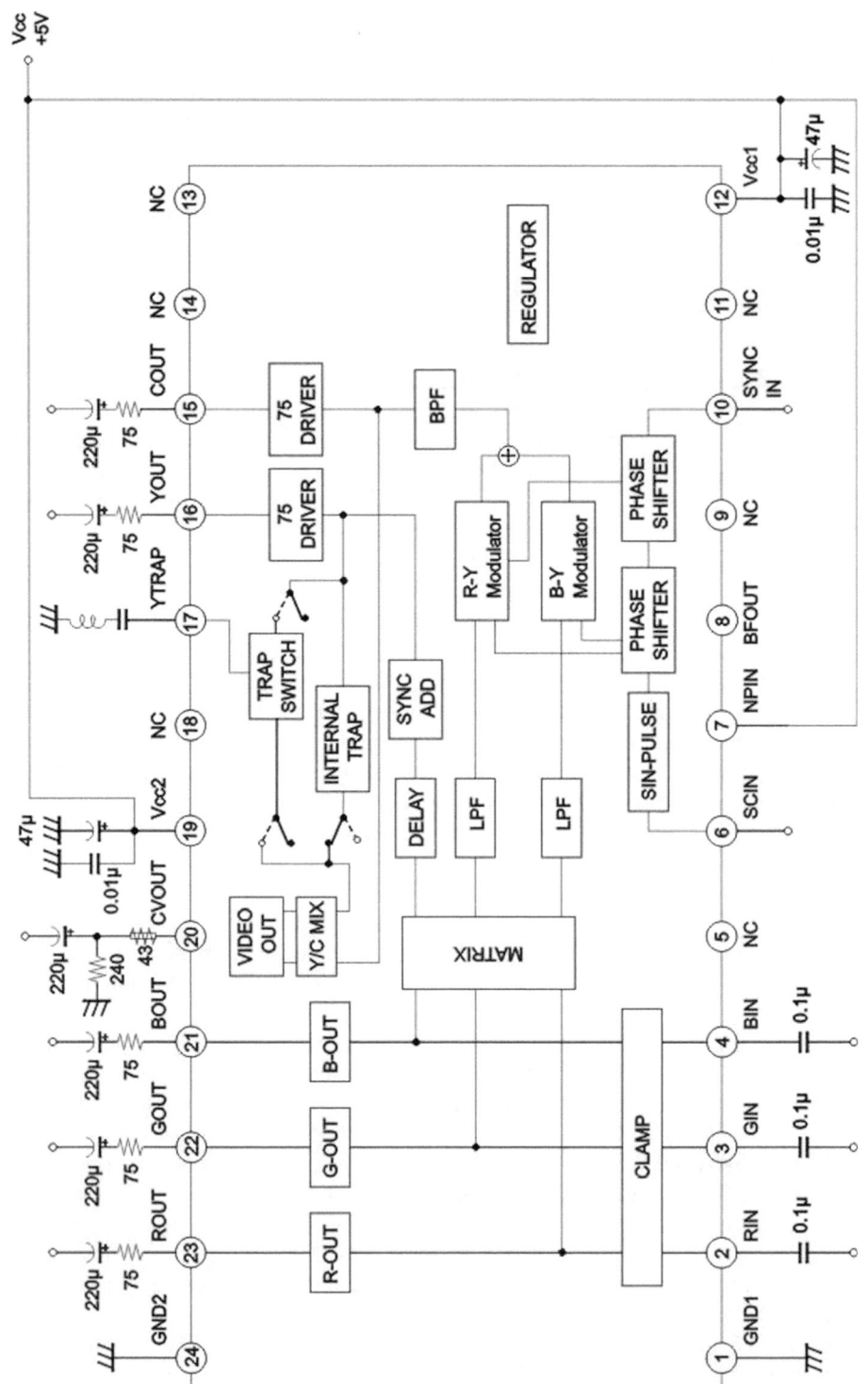

Fig. 27

3.7 Colour Encoders

V7040 is less well known than previous Sony chip encoders. It is described as an IC that can operate in both NTSC and PAL mode superimposing analogue RGB and outputs them as such, or as composite video signals. Both types of outputs can drive a 75 Ω load directly. For this reason it was chosen for Commodore's *A2300/A2301 Amiga Genlock*. In Fig. 28 there is the application note for PAL operation.

Analog Devices has a range of ICs suitable for video consumer and professional applications. It has produced four RGB analogue encoders, numbered sequentially: the *AD722*, *AD723*, *AD724* and *AD725*. Except for the AD722, they are still in production and Analog Devices can still offer samples on demand at the time of writing. Because they are of recent designs, these encoders offer excellent video quality as well as other features at low cost. Red, Green and Blue colour component signals are converted into their corresponding luminance (baseband amplitude) and chrominance (subcarrier amplitude and phase) signals in accordance with either NTSC or PAL standards. These two outputs are also combined to provide composite video output. All three outputs can simultaneously drive 75 Ω, reverse-terminated cables and all logical inputs are CMOS compatible. The chips operate from a single +5 V supply and no external delay lines or filters are required.

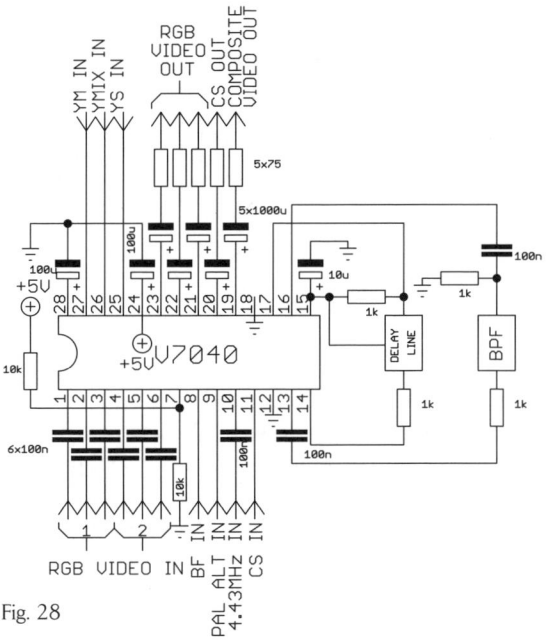
Fig. 28

The AD722 and AD724 accept either F_{sc} or $4F_{sc}$ clock. When a clock is not available, a low cost parallel-resonant subcarrier crystal and the AD722/AD724's on-chip oscillator generate the necessary subcarrier clock. The AD722/AD724 also accepts the subcarrier clock from an external video source while the other encoder chips only external accept F_{sc} subcarrier clocks. Some chips are well distributed from *Digikey, Farnell, Rochester Electronics, Arrow NAC, TME*, etc. In Fig. 29 there is an application note for the AD724 valid for the AD722 too.

Following the indications given for previous encoder chips, SPG625 can easily drive AD72x chips to create a very high quality video signal. The amplitude of RGB signals has to be lowered to 0.7 V to obtain full colour saturation and should be flat during the composite blanking interval. Internal circuitry will clamp this level during HSYNC to a reference that is used internally as the black level. The blanking level at the input pins can range between 0 V and 3 V with respect to the ground level. All the chips accept VS and HS separate sync inputs, but by setting them to logic levels '1' (+5V) or '0' (ground) the VS input, HS input can be used as composite sync inputs of negative or positive polarity, respectively. For more informations as well as to request free samples, please visit the Analog Devices website.

Also Philips produced an encoder chip, the *TDA8501* (24 pin DIP). It is described as a highly integrated PAL/NTSC encoder IC which is designed for use in all applications where R,

Chapter 3. Colour Standard Systems

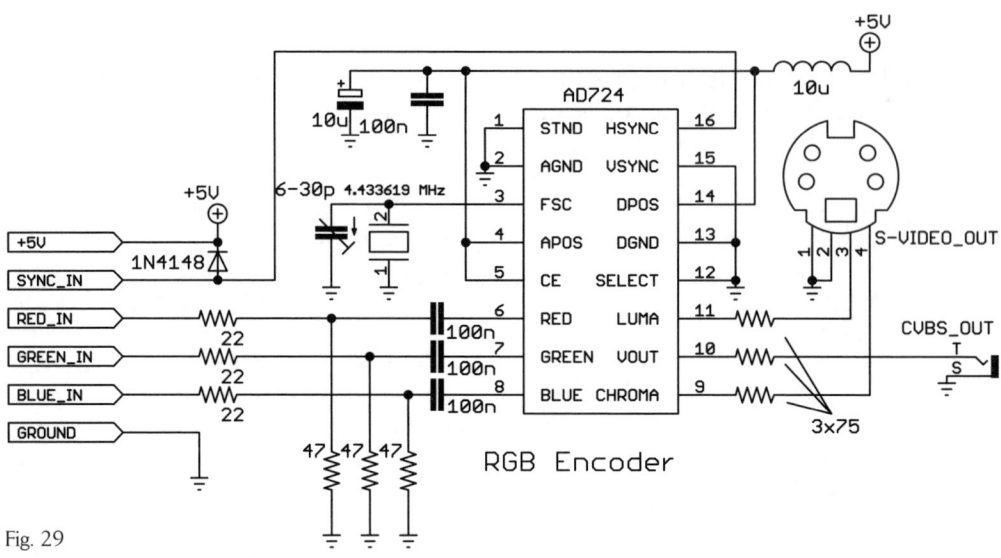

Fig. 29

G and B or Y, U and V signals require transformation to PAL or NTSC values. This IC owns all the features found in all the previous chips. The datasheet is freely downloadable from Philips website. In Fig. 30 there is a typical application note while the chip you can buy from TME (www.tme.eu).

Philips produced probably the sole RGB/YUV SECAM encoder chip, the TDA8505 (the block diagram is shown in Fig. 31). Following all the guidelines given to connect SPG625 in order to create a valid PAL/NTSC CVBS signal by means of the preceding PAL/NTSC encoder ICs, an expert technician could try to get an acceptable SECAM video signal from this device if also helped by the TDA8505's datasheets available from Philips website.

Again Philips produced a couple of SECAM encoder chips which could accept D_r and D_b signal as well as external synchronisms directly: they are the TDA2506 & TDA2507. Again furthermore information is available from a search on Philips' website or on the internet.

3.8 NTSC & PAL Colour Decoding

In principle, the process of decoding the colour signal follows a reverse process of the encoder. The system for the recovery of colour signals is called *Synchronous Demodulation*.

The composite video is a complex signal which needs to be recorded, transmitted, elaborated, etc. This as much as possible to be unaltered through the relative specific means as recorders, RF modulators, mixers, etc., to be decoded in the same initial RGB format and from which the pictures must be regenerated point-by-point as originally captured from a capture device.

The luminance can be easily extracted from composite video via a low pass filter with cut-off frequency of about 3 MHz for NTSC and 4 MHz for PAL and amplified as usually made in a B&W television. The luminance amplifier in a B&W TV set drives contrast and brightness circuits; in a colour TV set such circuits participate in the colour decoding process to obtain both B&W and coloured balanced images.

Composite synchronisms can be stripped from CVBS by means of one of the methods explained in the previous chapter and therefore the subject does not warrant further investigation.

In Figures 32 and 33 are illustrated the two possible block diagrams of a generic decoder for NTSC and PAL.

3.8 NTSC & PAL Colour Decoding

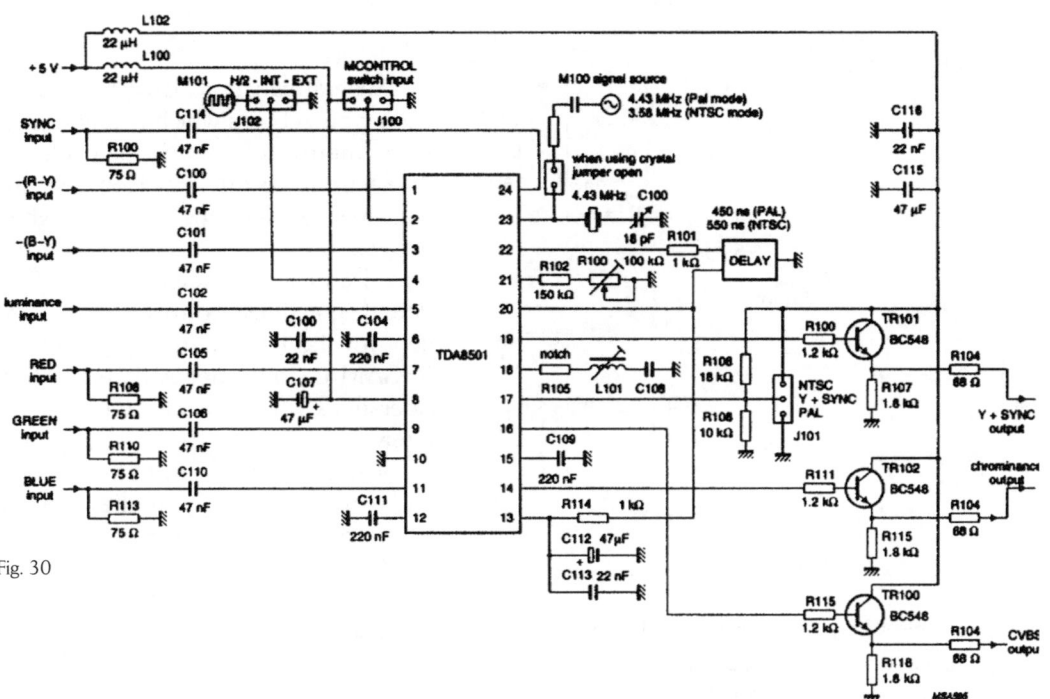

Fig. 30

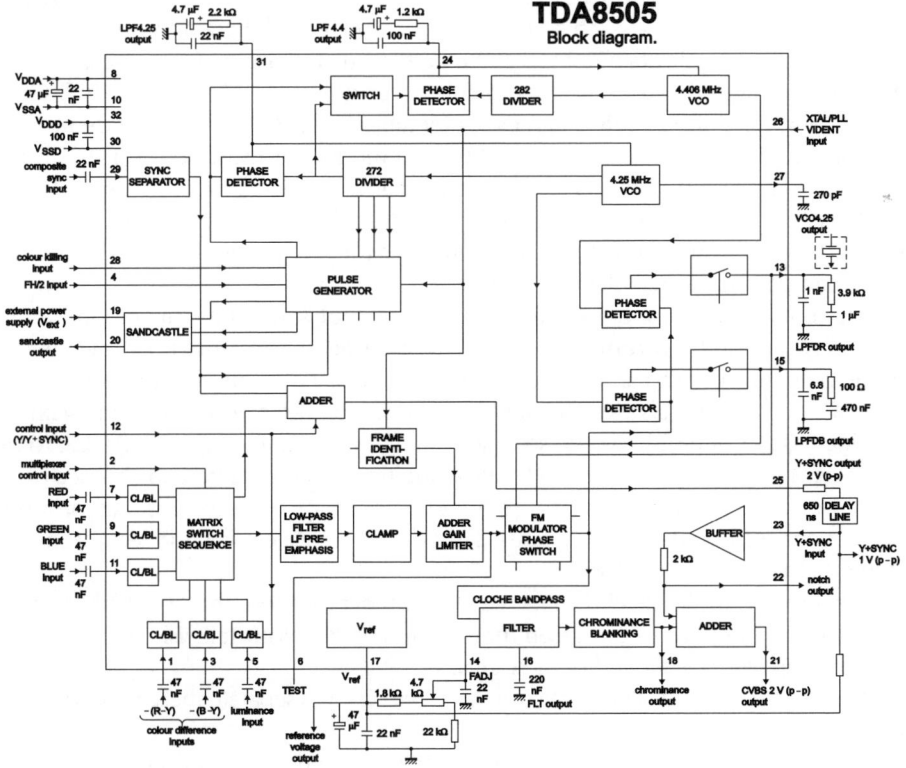

Fig. 31

Chapter 3. Colour Standard Systems

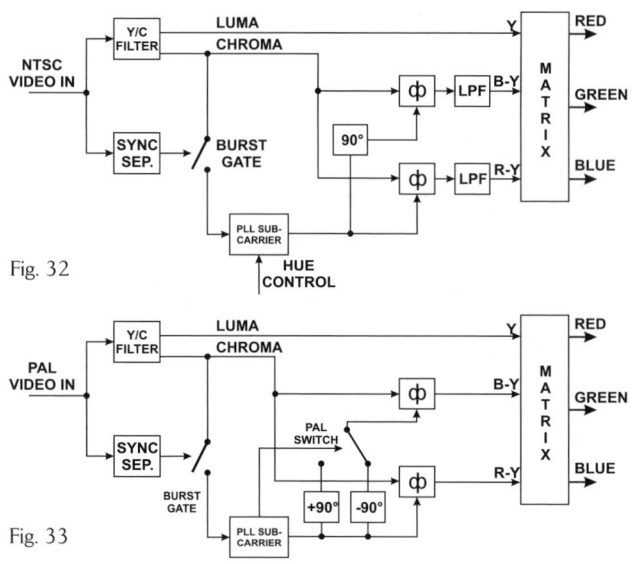

Fig. 32

Fig. 33

Several variations exist for the decoding procedures which over the years were improved following the evolution of electronic technology. However, the concepts remained the same through to today.

Essentially the technique for PAL decoding is similar to that of NTSC except for the line alternating process. A SECAM decoder needs different considerations which will be exposed separately.

The colour bursts must be revealed and extracted from composite video via a gate pulse recovered by a horizontal line flyback pulse. Next, a burst gate amplifier allows the bursts to pass through a *Phase-Locked Loop* (PLL) network known as an *Automatic Phase Control* (APC), which consists of the phase discriminator, a low pass filter and a *Crystal Voltage Controlled Oscillator* (XVCO).

For a more precise and stabile subcarrier oscillator, it was recently preferred to double or quadruple the XVCO frequency; the subcarrier clock is then obtained by dividing the output of the oscillator by 2 or 4. Such an oscillator is corrected in phase and frequency by a 'low-pass-filtered' error voltage obtained by comparing the burst phase with the oscillator itself by means of the phase discriminator. The locked subcarrier drives the V (or Q for NTSC) demodulator and of course is shifted of 90 degrees for feeding the U (or I for NTSC) demodulator.

For PAL only, the subcarrier phase for V demodulator must be reversed on alternate lines. In order to achieve this, the PAL burst signal carries a reversing of its phase line-to-line and a suitable circuit detects such phase flipping. The result is a half-line frequency square wave which will feed a PAL switching circuit. The techniques for recovering the two PAL or NTSC chrominance components are explained better below.

The burst is also responsible for the entire colour saturation calibration because the burst is the only part of the chrominance signal transmitted at constant and known level. Depending on the noise level or the amplitude of the received signal, alongside the comparison of level of syncs with level of bursts (which should be the same), the colour saturation gain can automatically increase or diminish via the so-called *Auto Chroma Control* circuit (ACC), to render 'noiseless' images with a suitable chroma level. In addition, if it is detected an absence of the subcarrier or the burst level is too low to cause a worse vision or even in presence of a non-compliant colour transmission, the entire colour demodulation process is inhibited through a suppressor circuit known as the *Colour Killer*.

Next step is to divide chrominance and luminance. Note that it is practically impossible to eliminate any interference between them and in early colour televisions simple filters based on coils and capacitors achieved poor results.

Later two improved filtering philosophies were used to separate these signals based on *notch* and *comb* filters.

The notch filter is used for the luminance component and its purpose is to cancel the carrier frequency of the chrominance in the composite video signal. The overall result obtained with this filter is similar to that obtained with a pass-band filter applied to the luminance signal. In fact, the filter starts to attenuate the luminance signal at frequencies slightly below the subcarrier and also has some benefit for frequencies slightly above the colour carrier, but the content of the luminance signal to these frequencies can be considered virtually nil.

The chrominance information in early days was separated from the luminance signal by a simple pass-band filter centred on the subcarrier frequency. In the mid '80s, digital technology became available to implement comb filters for chrominance filtering.

The comb filter is a special filter that adds to the signal a delayed version of the signal itself via a number of steps. The frequency response of a comb filter consists of a series of equidistant pulses that resemble the individual teeth of a comb. This device provides better separation between the two colour components and almost complete removal of cross colour effects.

Being stripped from luminance and via successive amplification, the chrominance enters into two balanced demodulators driven by two subcarriers, one phase shifted by 90 degrees respect the other one, to obtain, aided also by means of an add/subtract network, the weighted original colour difference signals, i.e. U & V for PAL or I & Q for NTSC. For PAL only, through a 64µS delay line, usually known as a *PAL delay line*, and a $F^H/2$ square wave pulse, the V component is phase-reversed line-by-line in order to remove the phase error. Only for the NTSC system, a manual I & Q phase calibration at this point is necessary to correct its notorious phase shift error.

Next and last task is to de-weight U & V, or I & Q for NTSC, to obtain R-Y and B-Y, and, via a complex matrix cross networking with luminance Y, to recover the original Red, Green and Blue colour signals. After buffering, these are ready to drive the relative CRT guns through a RGB power amplifier. Because of the colour decoding process, the chrominance is delayed with respect to the luminance and then a delay line is inserted into the luminance path to ensure that both signals arrive simultaneously to the final RGB matrix. The amount of this delay is fixed and it depends on the particularities of each design stage. It can typically vary from 200 to 800 nS. Excluding the radio wave propagation delay from the transmitter to the receiver, we can state that globally the images are reconstructed in our TV set screen from a minimum of 400 ns up to 1.6 µs delay time, values calculated by summing the delays of the encoder with the decoder.

It is necessary to DC clamp all three colour signals in order to anchor the black level of Red, Green and Blue signals to a common reference black level. Any DC drift level in the colour of one channel with respect to any of the other two will lead to unwanted colour dominances or, in worst cases, to variable and 'dancing' colours.

3.9 SECAM decoding

Similarly to PAL and NTSC, in SECAM the decoding process follows a reverse process of the encoder, hence the way with which luminance and synchronisms are stripped from CVBS and further elaborated.

The main difference from the other two standards is therefore the colour decoding process. The first step is to separate the chrominance signal from CVBS by means of a band pass filter generally centred at 4.43MHz ± 1MHz. The subcarrier signal is now subjected to the inverse of the *mise en forme* (form) feature, called the *cloche* (bell) circuit.

After this, the subcarrier is divided into two paths: one feeds one input of a commutat-

Chapter 3. Colour Standard Systems

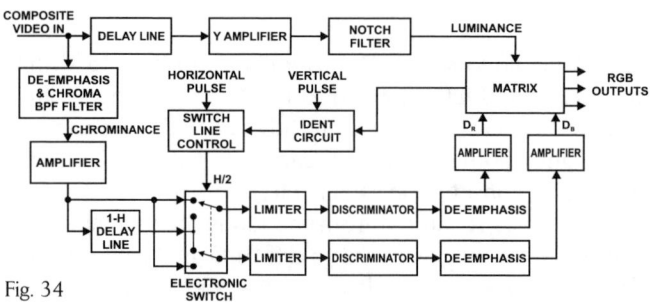

Fig. 34

ing switch, while the other path includes a 64µS ultrasonic delay line and an amplifier to compensate for the loss of delay line, terminating at the other input of the switch (Fig. 34).

The ultrasonic delay line device is very simple, i.e. a plate of quartz or glass medium and two transducers, usually made of barium titanate. Originally, the signal was introduced for 'bouncing' inside the shaped medium, later a straight rod was used with a transducer at both ends and, recently, this delicate component was inevitably replaced by a digital device based on CCD technology.

The intention is to recover from two subsequent lines that carry the information corresponding to the red difference component and then to the blue difference component, D_r and D_b respectively, a three-color image. The colour subcarriers on which the colour data travel are focused on different frequencies, namely 4.406 MHz (D_r) and 4.250 MHz (D_b), PLL-locked to the line sync revealer circuit. The SECAM decoding circuitry delivers D_r and D_b for the rest of the receiver to get RGB signals following standard colour television in practice equal to that of NTSC and PAL.

Both direct and delayed paths are made available simultaneously to the decoder output, despite the transmission of colour difference signals in sequence, although the vertical resolution in colour is almost halved. During the second active line, we assume that D_r was sent, and therefore, by definition, in succession, the line was an earlier version of D_r line. Then at line 2, D_r is emerging from the delay line. Therefore, at the inputs to the commutating switch, the two difference signals are present together. During the third line, D_b is being transmitted and D_r emerging from delay line.

It remains only to consider the circuit that restores the commutating switch to the correct sequence. The earliest SECAM version inserted the identification signal into vertical blanking with opposite polarity for red and blue difference signals. When special signals were inserted inside the same space, such as teletext, this technique was subsequently abandoned in favour of a 'colour' burst pulse, carried line by line as in the PAL and NTSC systems, thus ensuring a better switching performance. The switch is actuated by a flip-flop circuit which is activated by pulses from the receiver line time base and reset in the right sequence by the identification pulse.

With the operation of the commutating switch at the end of each line during blanking period, each colour difference signal is fed to its appropriate limiter and discriminator.

The resulting D_r and D_b signal are finally scaled and recombined with luminance Y to give Red and Blue signals. The Green signal is obtained by a simple subtraction of the difference between R and B from Y. Finally, the three RGB signals are ready to drive the CRT guns as usually done in regular TV colour systems.

The SECAM system is able to give good colour pictures and has the merit of being insensitive to differential gain and phase distortion in the path of transmission. Despite the apparent complication of the delay line, the receiver is otherwise simple and uses standard techniques. Also with SECAM, for the first time European viewers easily could re-create good colour television images in their houses.

The above description covers just one of the numerous SECAM variants, i.e. the oldest one. Successive versions brought improvements and some incompatibilities between them. The above discussion is still useful to provide an idea about how the SECAM decoding procedure works, generally applicable to all variants.

3.10 Colour decoders chips

Just as for the encoders, the first colour decoders were made of bulky discrete components and electronic tubes. Later the electronics moved to discrete solid-state components and further to integrated circuits, miniaturizing off the television sets. Since the decoders were destined for use in the volume market of home televisions, the manufacturers of electronic components ended up producing almost infinite numbers of dedicated ICs all designed for decoding the colour composite video signal in a TVC set. Some semiconductor producers turned out also bi & tri-system decoder ICs, and even multisystem ICs capable of decoding practically any colour sub-standard available across the world.

It is practically impossible to list here all these ICs!

Hence, we shall itemize only some popular and interesting ones used in commercial televisions up to today. Some ICs are even able to convert pre-digitalized video data in YUV/RGB component signals for further processing or to be displayed on a screen. Other devices could perform a rough conversion between standards.

3.10.1 Analogue Decoders

Substantially good RGB signals from CVBS in the early colour TV sets were effected using such popular ICs as *TBA520, TBA560 & TBA540* (this last often coupled with the PAL/SECAM *TCA640* or *TCA650*, or the improved PAL-only chroma demodulator *TBA990*), or the other triplet *TDA2140, TDA2151 & TDA2161*, or the multistandard decoders *TDA4555* and *TDA4556*. Over the years improvements were progressively achieved to minimize the relative colour standard defects. So while for NTSC decoders, comb filters and an automatic hue control (calibrated by default on a flesh/skin tint) were embedded, for standard PAL decoders a rather fancy (read expensive) 1-H glass or quartz delay line was employed. Some TVC manufacturers, for example Sony, in targeting cheap PAL TV sets, had initially preferred to adapt existing NTSC decoders to create a simple decoding procedure by adding a user accessible external control for 'hue' control so to avoid the necessity of this expensive 1-H delay line.

Similar to piezoelectric converters, due to mass production and further miniaturization, the 1-H delay line function became truly inexpensive to manufactures. Ultimately the 1-H fragile ultrasonic glass delay line was incorporated into a single device based on CCD technology whose single internal elements were employed to shift the transportation of analogue signals (in this case electric charges) through successive stages (capacitors), controlled by a clock signal, to form a delay line. In this way, it was shown possible to delay, demodulate, and matrix the chroma signal from the chrominance filter all in the charge domain in one complex CCD with phases according to the colour subcarrier. The output signals could be fed directly to the RGB matrix as usual to drive the RGB CRT guns. Thus the whole chroma section of a PAL TV set now consisted only of one CCD IC performing the colour signal processing plus one or two bipolar ICs with the 4.43 MHz oscillator, PAL switch, etc., thus eliminating the external glass delay line and coils.

The use of this particular solid-state component created an efficient chroma demodulation

Chapter 3. Colour Standard Systems

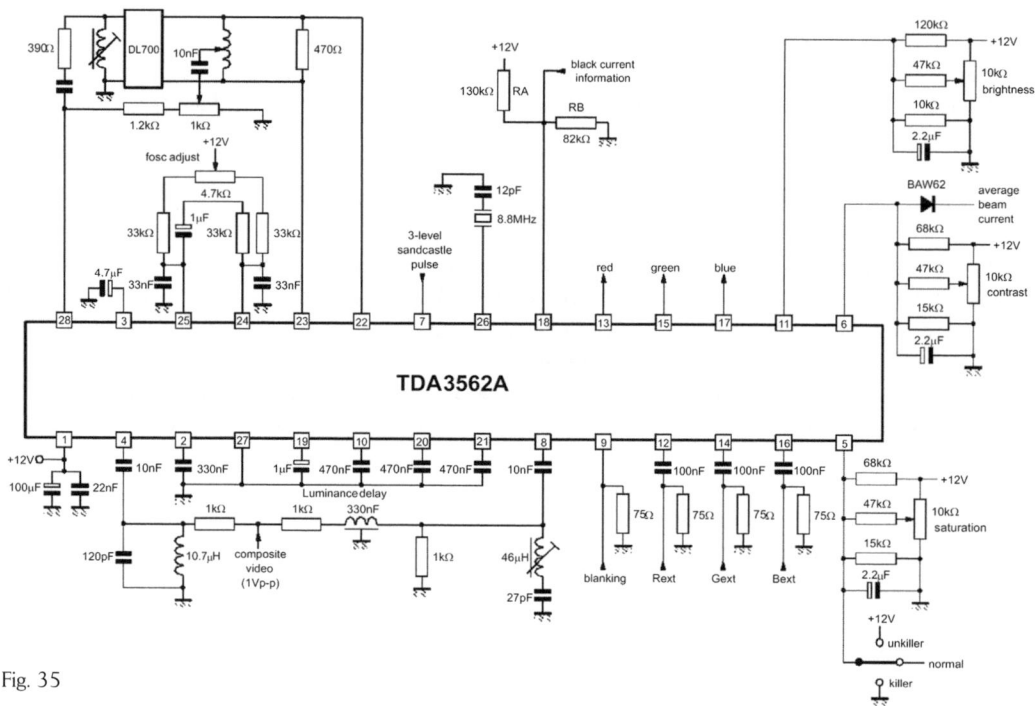

Fig. 35

virtually with no difference of gain for the line-by-line alternating V component because the CCD IC, being an active electronic component, could manage to compensate differential gain. The *TDA4660* (from Philips) and *STV2180* (from ST Microelectronics) are two examples of CCD base band chroma delay line ICs.

One of the most popular bipolar ICs capable of extracting RGB signals from CVBS (or S-Video) was the *TDA3562A* (produced by Philips, ST and other manufactures) whose application note in Fig. 35 shows its design simplicity. The DL700 is the glass ultrasonic 1-H delay line.

The *TDA3565* is a 'reduced' version of the TDA3562 from which was eliminated the RGB external inputs needed for Teletext and OSD messages. Note that for the correct functioning of all these devices a special multilevel signal was introduced, which for its particular shape was called a _Sandcastle Pulse_ (SCP), or even a _Super Sandcastle_ pulse (SSC), consisting of various mixed simple pulses arranged at different voltage levels, such as the Burst Gate pulse, Horizontal Blanking and Vertical Blanking. Internal voltage comparators inside the TDA3565 (or equivalent device) extracted these pulses needed for the correct decoding procedure. The sandcastle pulses are specific for TV colour circuits and are created by specialized ICs, such as the *TDA2595*, *TDA 2579*, *TEA2130*, etc., which could also perform horizontal & vertical synchronization. The use of the sandcastle pulse also simplified the PCB design because with one track three signals are carried simultaneously.

In Fig. 36, 37 and 38 there are the schematics of a complete PAL-only CVBS/S-Video RGB Decoder. This unit could be used to give a composite or separate Y/C video input to PAL video displays furnished with only RGB inputs. The relative complexity of the circuit is counterbalanced by the extreme straightforwardness of its calibration: just one adjustment performed and without any specific instrumentation!

The circuit was divided into three parts for assisting in description of the subsystems. Part 1

3.10 Colour decoders chips

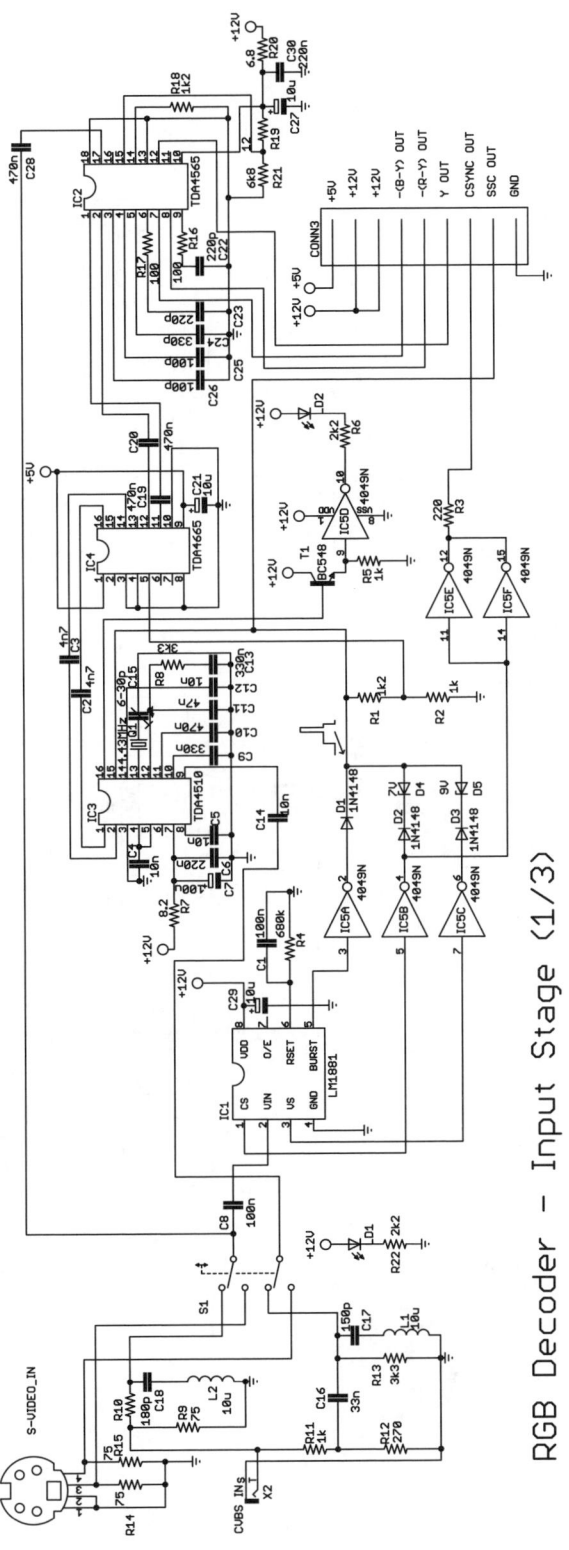

Fig. 36

Chapter 3. Colour Standard Systems

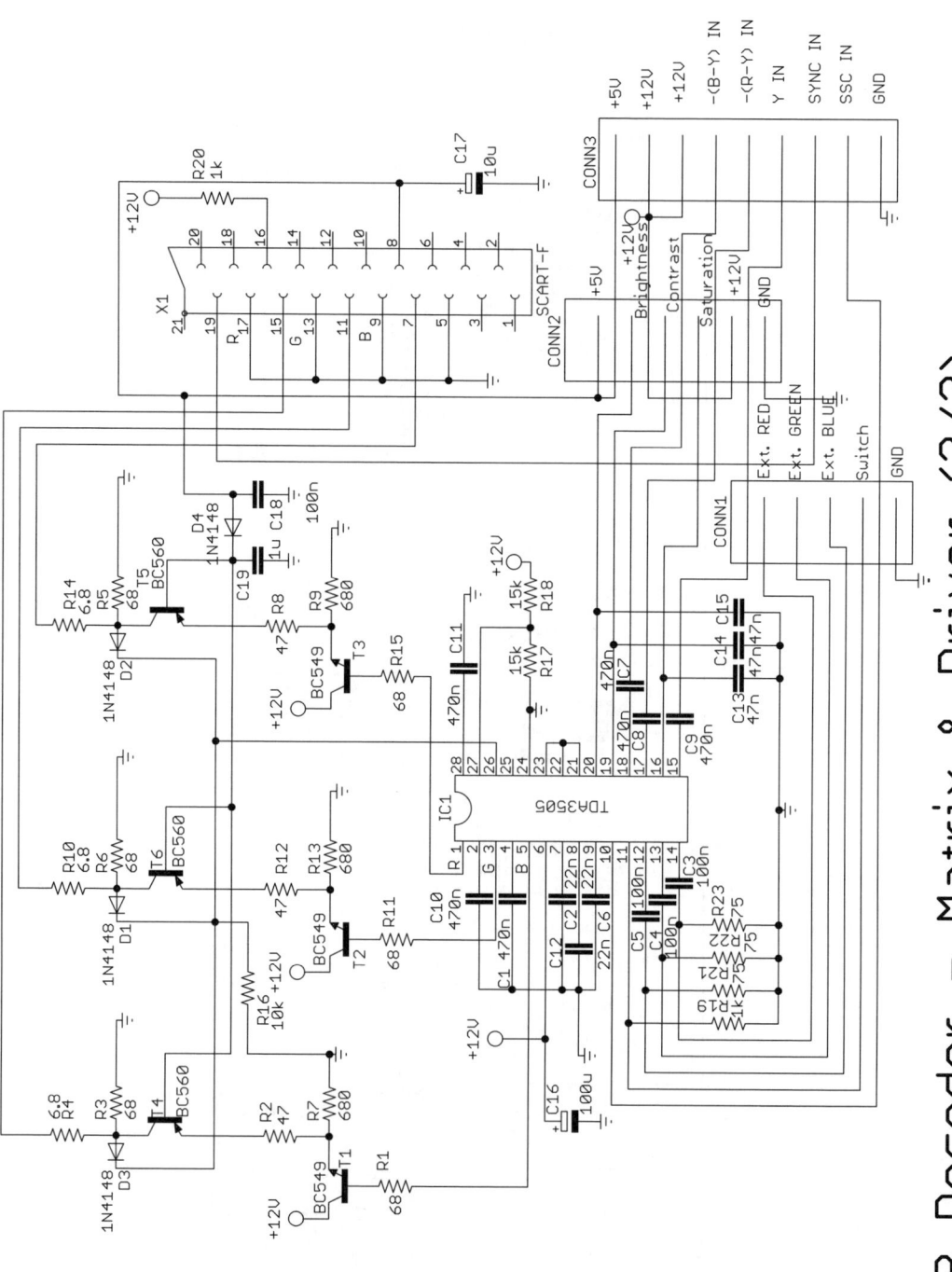

Fig. 37

3.10 Colour decoders chips

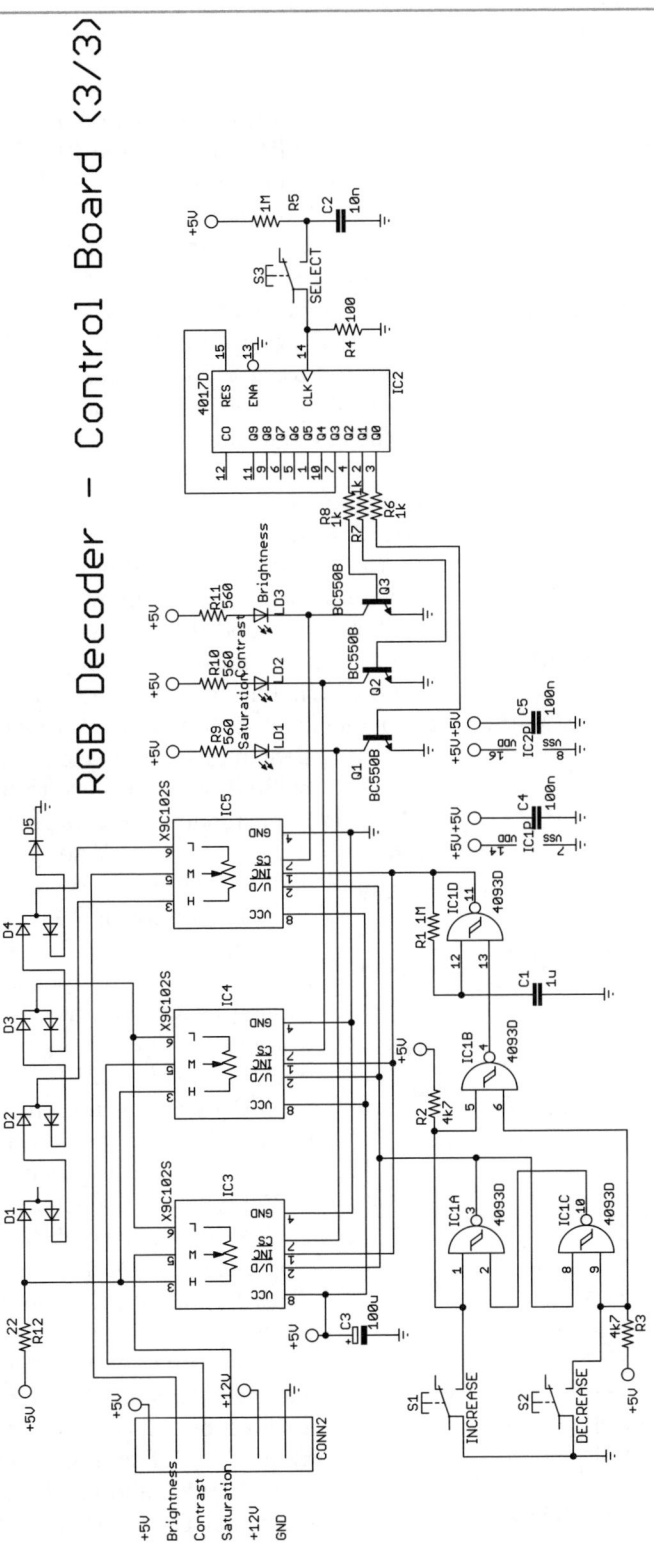

Fig. 38

Chapter 3. Colour Standard Systems

(sheet 1/3, Fig. 36) regards the input stage, which includes the chroma/luma filters, the sync separator, the super sandcastle pulse generator, the PAL decoder, the 1-H CCD delay line, the <u>C</u>olour <u>T</u>ransient <u>I</u>mprovement circuit (CTI) and the luminance delay line. Part 2 (sheet 2/3, Fig. 37) converts Y/B-Y/R-Y signals in true RGB signals to feed a SCART output whilst the remaining part 3 (sheet 3/3, Fig. 38) concerns the picture brightness, contrast and saturation level controls set digitally via three electronic potentiometers.

The networks made up of R10/C18/L2 and R11/R12/C16/R13/C17/L1 are the filters for separating luma and chroma from CVBS. Y & C signals from the S-Video IN socket are connected to S1 without the CVBS source selected using S1, a dual 1way/2 position switch. S1 performs a 'switch mode function' to display alternatively a CVBS or S-Video source without connecting and reconnecting the relative cables originating from the same or different equipment.

It was preferred to use standard RLC filters in place of a comb filter IC (as for example the Philips *TDA9181*) because the latter needs a subcarrier-locked external clock frequency to operate correctly and the relative circuitry to achieve this is too complex to build. However in Fig. 39 the TDA9181's application note circuit could help to employ this comb filter chip by adapting it to our RGB Decoder. More informations and datasheets are available on Philips' website to assist designs with an appropriate subcarrier locking system.

Luma & Chroma signals travel through two different paths. Luma feeds the sync separator IC1 (LM1881) and IC2 (TDA4565), a monolithic integrated circuit designed for colour transient improvement and luminance delay line, realized using a gyrator technique. A specific input for luminance signal path substitutes the conventional Y-delay coil with an integrated Y-delay line, switchable from 730 to 1000 ns in steps of 90 ns and an additional fine adjustment of 50 ns by setting pin 13 to ground. Two Y output signals, one of 180 ns less delayed, are available too. The control voltage applied to pin 15, derived from potential divider R21-R19, sets the Y delay to 780 ns in the present circuit providing finally a delayed luma signal from pin 12. This delayed luma signal is also connected to CONN3 to be used later.

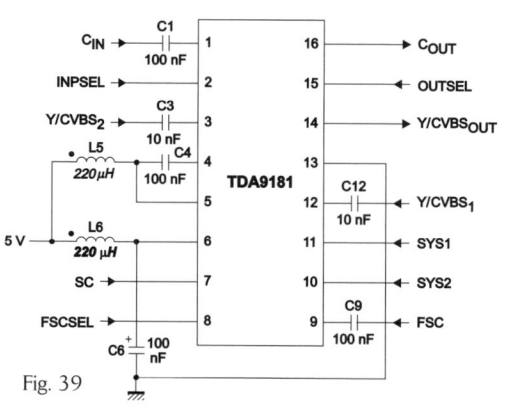

Fig. 39

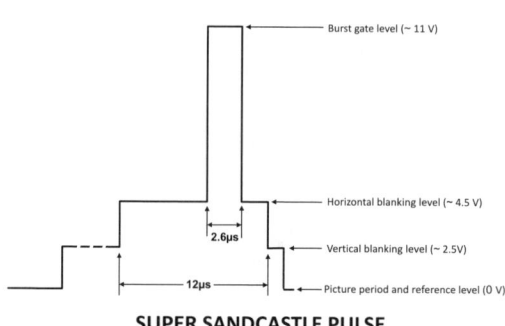

Fig. 40 — SUPER SANDCASTLE PULSE

The luminance from S1 also supplies IC1 which extracts the burst gate pulse (pin 5, BURST), the vertical (pin 3, VS) and composite syncs (pin 1, CS). To create the necessary sandcastle pulse having the correct form and level steps, BURST, CS and HS are first inverted by IC5A, IC5B and IC5C, respectively, three of six buffer/inverters included in a 4049 IC, and via the diodes D1 to D5 (note that D4 & D5 are zener diodes) and

3.10 Colour decoders chips

R1+R2. The correct voltage level steps are then generated with the relevant timing information embedded: 0 V (ground) = picture period and reference level; from 2.15 to 3 V = vertical blanking level; from 4.1 to 4.5 V = horizontal blanking level; >10V = burst gate level (Fig. 40). The sandcastle pulse (SSC) so created feeds IC3 (TDA4510) and IC4 (TDA4665). Successively this pulse will be employed to drive the video control functions IC1 as shown in part 2 (sheet 2/3) which will be analyzed later.

IC3 (TDA4510) is at the heart of the PAL decoding. By simply inputting a chroma signal at pin 9 and a sandcastle pulse at pin 15, aided by a $2F_{sc}$ PAL quartz crystal Q1 and a few passive components, -(R-Y) and -(B-Y) are obtained from pins 1 & 2, respectively. A high level at pin 16 indicates the correct PAL decoding: through T1, a level translator, and the inverter/buffer IC5D, the LED LD2 is illuminated.

Next step is to create the 1H-delays for the colour difference signals. To achieve this, IC4 (TDA4665), a CCD base band chroma delay line, is used which is basically a solid-state component able to store a 1-H line for PAL or SECAM systems. A 3 MHz internal clock signal is derived from a 6 MHz clock line-locked by the sandcastle pulse inserted (by halving its voltage) at its pin 5. This clock drives the internal CCD analogue shift register obtaining a precise 64µS retard with no extra passive or active components, becoming *de facto* a welcome alternative to the 'old' glass delay line and eliminating at the same time the complex phase & amplitude adjustments.

The delayed -(R-Y) and -(B-Y) signals pass via C19-C20 through IC2, which we previously have employed as a luma delay line, but in this case it does another useful function of improving colour pictures. Its internal colour transient improvement circuitry is able to detect a colour transient by differentiating the colour difference signals, performed by an internal difference amplifier and C25 & C26. When a transient is detected, an internal pulse shaper is actuated being based on the value of C24. The pulse shaper causes the input signal to be stored in a *Sample-And-Hold* (S&H) circuit, which retains the current signal level until the transient has passed. Next, 100 ns later the new level is supplied. The S&H function is implemented by external components R17, R16, C23 and C22. The re-shaped colour difference signals at output pins 7 and 8 of IC2 are fed to the matrix circuit (sheet 2/3) via CONN3.

The remaining inverter/buffers IC5E-IC5F are used for amplifying and buffering of composite synchronisms to feed CONN3 and later pin 19 of a SCART connector (see sheet 2/3). LD1 is the power-on lamp.

The colour matrix circuit (sheet 2/3, Fig. 37) is based on the TDA3505 in which the luminance and chrominance signals meet. The basic colours R, G and B are recovered by a summary matrixing operation of colour difference signals and luminance Y component. The image settings (contrast, brightness and colour saturation) are driven directly by voltages that determine bias and gain at a certain number of points in the matrix. Two-stage level shifter/buffers are needed at the matrix outputs, because these levels do not go down to 0V and are not able to handle a 75 Ω load directly. The buffering and level shifting are performed with three combinations of an emitter follower and a common base amplifier (T1-T6). The output impedance of the three drivers is 75 Ω.

Each colour output driver has a diode which allows the operating point of the two-transistor stage to be monitored via pin 26 of the TDA3505. The operating point is monitored and, if necessary, corrected during the vertical blanking interval. The direct voltage required for this function is stored in capacitors C1, C10 and C11 during the actual picture. The matrix circuit

Chapter 3. Colour Standard Systems

recognises the vertical blanking period with the aid of the SSC pulse present at its pin 10.

To adjust the brightness, the contrast and the colour saturation, I designed an elegant tri-function digital potentiometer able to furnish and store a specific voltage level needed for each adjustment. However these voltages are not the same for the three regulations, and therefore the problem must be solved differently if we want to use electronic potentiometers in the place of standard passive 1-turn manually preset potentiometers. In Fig. 38 we can see the control board schematic based on 5 ICs and few other components.

IC3, IC4 and IC5 (X9C102S) are 1kΩ/100-steps digital potentiometers provided from Intersil and available from Farnell or Digikey. They can be substituted by the cheaper Intersil's X9313Z, the 32-steps version. With this solid-state potentiometer, it is possible via simple digital pulses to control a resistor array composed of some resistive elements and a wiper-switching network to make up a digitally controlled three-terminal potentiometer. Between each element and at either end the tap points (H & L) are accessible to the wiper terminal. The position of the wiper element is controlled by the CS, U/D, and INC inputs and can be stored in a non-volatile memory, which can be recalled upon a subsequent power-up operation.

Three push-buttons are sufficient for a full calibration; two buttons (S1 & S2) are used to increase and decrease the voltage calibration needed by TDA3505 and the last one (S3) selects the regulation. IC1A & IC1C (two of four 2-Input NAND Schmitt Trigger included in 4093) form a S/R Flip Flop driven by the logic state at pin 1 and 9 connected to the correspondent S1 & S2 push buttons.

By depressing S1, a '0' logic level occurs at pin 1 of IC1A placing a '1' on pin 3 which will set the direction of the wiper movement through the logic level at pin U/D: a '1' enables the movement of W 'cursor' towards pin H otherwise to pin L if we press the S2 button, resetting the flip flop also. The S1 or S2 actions affect also the logic state of the third NAND Schmitt Trigger IC1B that enables the simple oscillator R1, C1 and IC1D, the last NAND Schmitt Trigger. A single pulse from IC1B changes (just once) the logic level at pin 11 of IC1D whilst, by pressing and holding down S1 or S2, IC1D starts to oscillate creating a clock whose frequency is set by R1/C1 values. This clock is used to increment or decrement the counter in the direction indicated by the logic level on the U/D input. The frequency of the clock can be changed empirically to match the optimal speed for a full calibration range, according to the number of steps chosen for the potentiometer, by varying R1 or C1 value.

Because U/D and INC pins of all the digital potentiometers are driven in parallel, it was necessary to select which potentiometer has to be enabled to singly adjust brightness, contrast or saturation. An appropriate '0' logic level at pin CS of either potentiometer is selected by IC2, a decade counter type 4017, using successive S3 push-button de-presses and through Q1, Q2 and Q3 (used as inverters and LED indicator drivers), enabling one-by-one each potentiometer.

Diodes D1 to D5 (note: D1 to D4 are dual diodes which can be replaced by standard single diodes type 1N4148) create a 'voltage-stepper' having 0.5V for each step which created the necessary reference voltages to fully control the RGB Decoder regulations. In particular IC3 calibrates the colour saturation by a voltage range from 2 to 4.5V, IC4 the contrast (2 to 4.5V) and IC5 the brightness (1 to 3V).

Whosoever may wish to employ some front-panel mounted traditional potentiometers may, of course, replace the digitally controlled counterparts thus to give a continuous control range rather than a step-by-step. The relative schematic is visible in Fig. 41.

This digital control board can be used for other purposes, such as for example to mix audio

3.10 Colour decoders chips

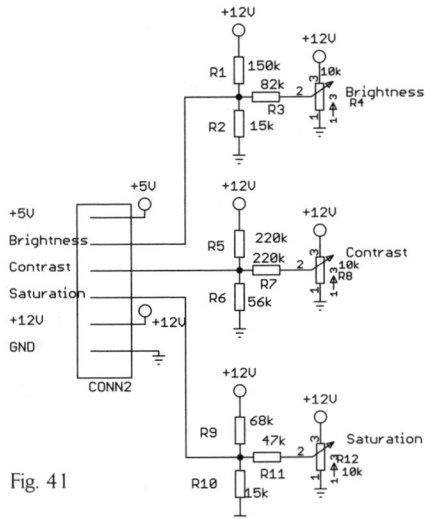

Fig. 41

signals, taking into account that there are some limitations. Being based on digital CMOS chips supplied by +5VDC, it is not possible to swing fully to ± 5V referred to ground for adjustments and the maximum current tolerable by each potentiometer IC is about ±4.5 mA. Another limitation is that there is no visual evidence of the exact wiper position, except for the initial power-on set at half way by default. However, their high reliability, the endurance of 100,000 data changes per bit, and over 100 years of register data retention, make these ICs a valid, cheap and modern alternative to 'old' mechanical potentiometers. Other suggestions and informations concerning these interesting devices are available on the Intersil website.

The SCART socket that supplies the RGB output signals also carries the AV and SWITCH voltages (+12V and +5V for automatic switchover to AV and RGB mode respectively).

CONN1 switches between internal and external RGB signals, using a TTL logic level '0' or '1' respectively at its 'Switch' pin, for displaying OSD messages or Teletext captions.

The RGB Decoder needs two regulated power supplies (+5V and +12V, at about 400 mA in total) applied in all the respective pin +12V and +5V of CONN2 and CONN3 connectors.

The calibration phase is very simple. Insert a PAL CVBS or S-Video signal at CVBS or Y/C input socket and switch S1 (sheet 1/3) accordingly: a greyscale image will appear on the screen display. Slowly regulate C15 to get a correct coloured image: LED LD2 will turn on. Next, set C15 in a half way between the two colour switch-off positions across its full rotation. Then push S1, S2 and S3 (sheet 3/3) to modify brightness, contrast and saturation according to your own individual preferences. That's all!

If by rotating C15 in a complete revolution a greyscale image is still present at RGB outputs as well as LD2 remains turned off, check using an oscilloscope if the sandcastle pulse is present and has the correct shape and levels, as explained previously, both at pin 15 of TDA4510 and (halved) pin 5 of TDA4665 so as at pin 10 of TDA3505. If any pulse is not available, check IC1 (sheet 1/3) and the components around it.

If a *moiré* pattern is too evident on the screen, it can be attenuated by inserting in parallel to C1 (sheet 1/3) a small 10-60 pF variable capacitor and then adjusting for minimum *moiré* interference in the colour picture.

3.10.2 SECAM decoders

Generally SECAM decoder ICs are used in conjunction with PAL decoding chipsets to take advantage of the common 1-H delay line, matrix combination and RGB CRT gun drivers. In practice the purpose of the SECAM section in a TV set is only to recover B-Y and R-Y from SECAM CVBS delivering the PAL counterparts to do the rest of the task. One of these SECAM ICs is the Philips's TDA8395, a self-calibrating and fully integrated SECAM device albeit it should preferably be used in conjunction with the PAL/NTSC decoder TDA8362 or TDA8366 and with the CCD baseband delay integrated circuit TDA4660. The TDA8395 incorporates high and low frequency filters, a demodulator and an identification circuit (luminance is not

Chapter 3. Colour Standard Systems

processed in this IC). The IC needs no adjustments and very few external components are required. A highly stable reference frequency is required for calibration and a two-level sandcastle pulse for blanking and burst gating (in Fig. 42 its block diagram).

The TDA8490 is an old monolithic integrated SECAM decoder. This circuit is intended to be used in conjunction with TDA8390 or TDA8461 (PAL decoder), TDA8451 (delay) and

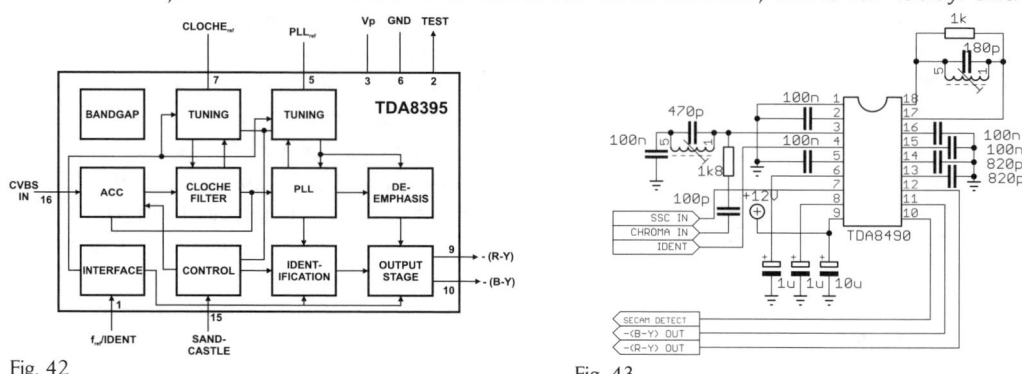

Fig. 42

Fig. 43

TDA8452 (filter). In the application note example shown in Fig. 43 the TDA8490 is implicit to be placed in parallel with the demodulation circuit of the PAL decoder. The internal identification circuit is able to detect both horizontal and vertical SECAM identification pulses.

We may underline also the Sony's CXA1214P, a SECAM signal processing IC which, with the combined use of CXA1213S, made it possible to configure a system compatible with all three systems, PAL, SECAM and NTSC. It had a self-contained automatic ID determination circuit (Fig. 44).

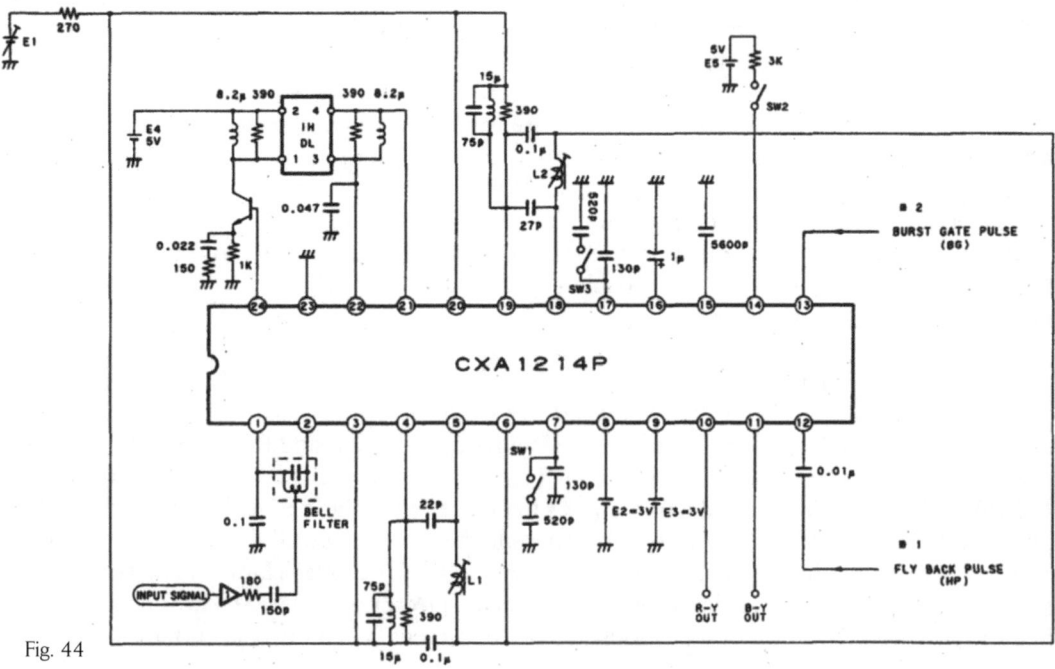

Fig. 44

3.10 Colour decoders chips

3.10.3 Digital & Pseudo-Digital Decoders

Since the first MOS integrated circuits were used in colour receivers in 1973, TV industries began to examine the possibilities of applying digital technology on large scales in the television set sector. Since 1977 some companies increased their efforts in R&D, whose results led to the decision to begin the digitisation of TV receivers.

Furthermore, to propose on the market cheap models of TV sets capable of displaying images in any standard, regardless of the country where they would be destined or sold, many multistandard system ICs were designed and produced by global semiconductor brands, as for example Philips Semiconductors which proposed also some chips based upon a special own-designed 2-lines-data communication protocol called *I²C* opening a large path for the age of fully digitally-controlled adjustments of TV sets instead of manual calibrations effected still in those times.

In the following years, the I²C protocol became a standard *de facto*. I²C is quite complex and a simple, working I²C coding circuitry is hard to implement using standard ICs to drive the relative devices (for which were designed some simple microprocessors programmed by basic software). We will examine only summarily some popular I²C chip typical examples without proposing working circuits.

The various versions of the *TDA935X/6X/8X*, produced by Philips itself, combine the functions of TV signal processor together with a µ-Controller and a US closed caption decoder. Most versions have on board a Teletext decoder which has an internal RAM memory capable of storing from 1 to 10 pages of text, depending on the model (Fig. 45 & 46). More information is presently available on Philips website.

The *STV224XH/228XH/223XH*, produced by ST Microelectronics, are fully bus-controlled ICs for TV that include <u>P</u>icture <u>I</u>ntermediate <u>F</u>requency (PIF), <u>S</u>ound <u>I</u>ntermediate <u>F</u>requency (SIF), Luma, Chroma and Deflection processing functions. Used with a vertical frame booster (such as the TDA8174A or the STV9306), they allow multistandard sets to be designed with very few external components and no manual adjustments (Fig. 47). Of course, other detailed information is available on ST Microelectronics website.

The *TVP5146* from Texas Instruments is a NTSC/PAL/SECAM, 4x10-Bit Digital Video Decoder with APS (*Analogue Protection System*) Macrovision™ *Detection*, YUV/RGB Inputs, 5-Line Comb Filter and SCART Support device.

This single-chip high-quality digital video decoder digitizes and decodes all popular baseband analogue video formats into digital component video. The TVP5146 decoder supports the analogue-to-digital (A/D) conversion of component RGB and YUV signals, as well as the A/D conversion and decoding of NTSC, PAL, and SECAM composite and S-video into component YUV. This decoder includes four 10-bit 30-MSPS (<u>M</u>ega <u>S</u>amples <u>P</u>er Second) A/D Converters (ADCs). Preceding each ADC in the device, the corresponding analogue channel contains an analogue circuit that clamps the input to a reference voltage and applies a programmable gain and offset. A total of 10 video input terminals can be configured to a combination of

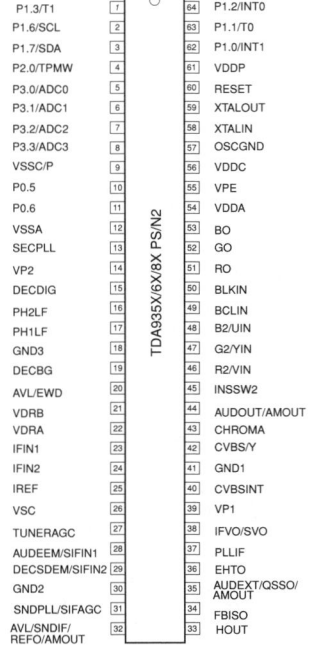

Fig. 45

Chapter 3. Colour Standard Systems

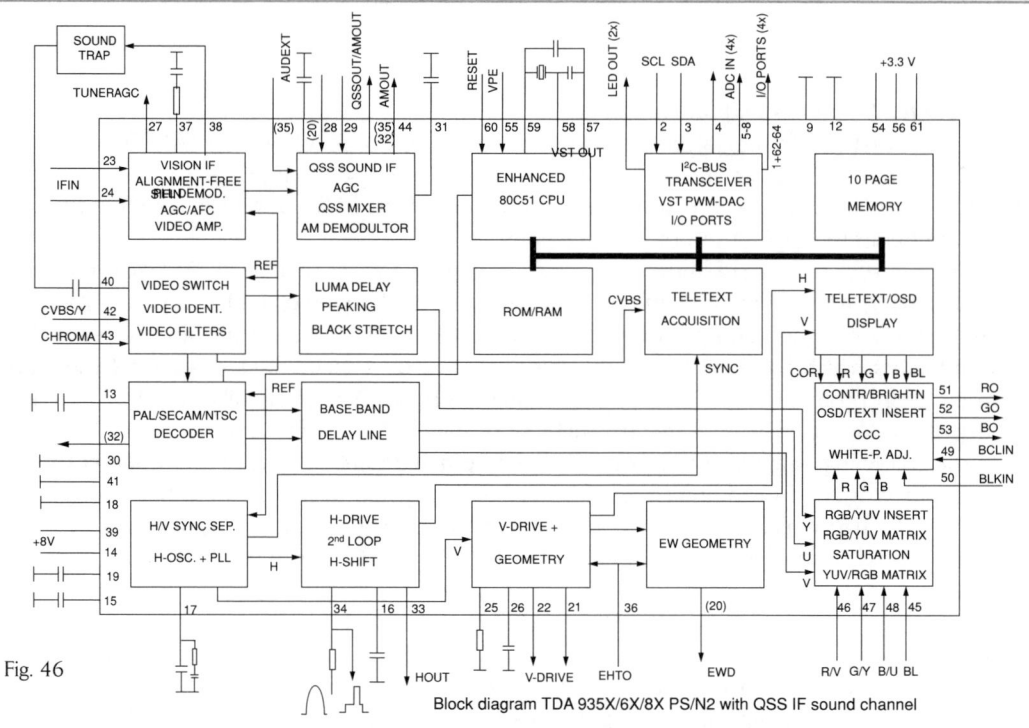

Fig. 46 Block diagram TDA 935X/6X/8X PS/N2 with QSS IF sound channel

Fig. 47 STV2238H — MULTI-STANDARD EUROPEAN APPLICATION PAL/SECAM/NTSC BG/DK/I/LL'

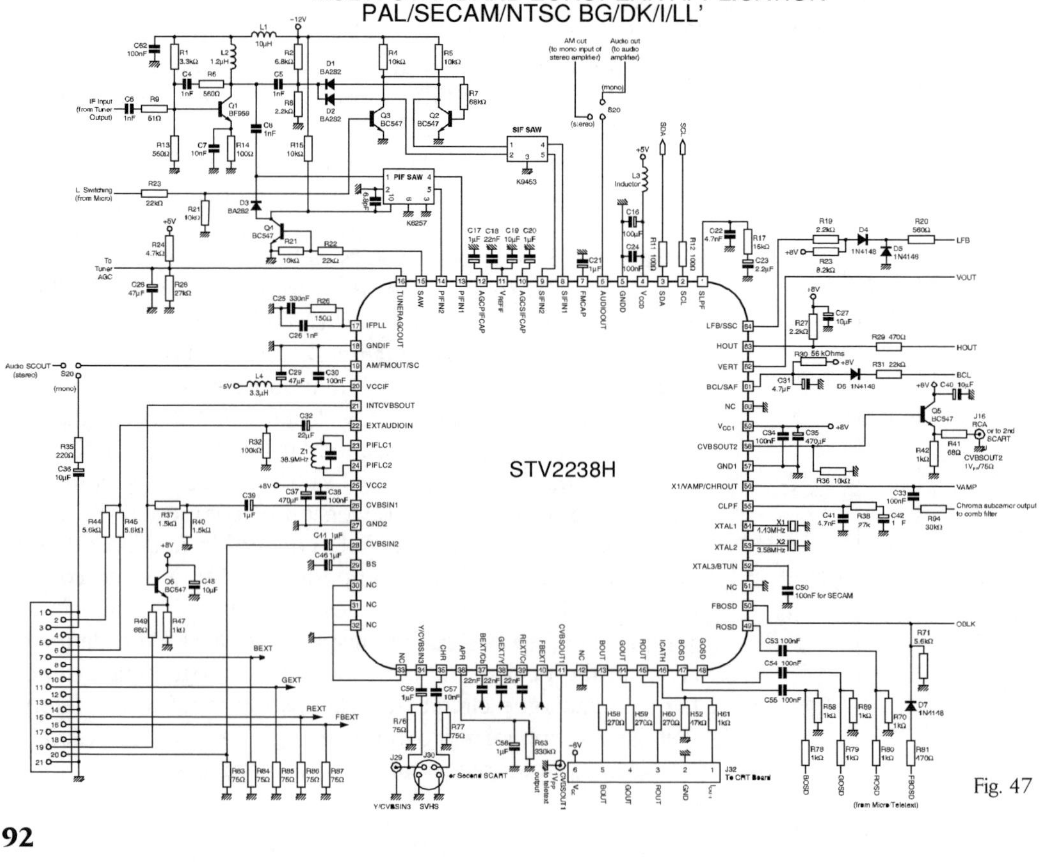

3.10 Colour decoders chips

RGB, YUV, CVBS, or S-video video inputs.

Component, composite, or S-video signals are sampled at 2x line-locked square-pixel clock frequency, and are then decimated to the 1x pixel rate. CVBS decoding utilizes five-line adaptive comb filtering for both the luma and chroma data paths to reduce both cross-luma and cross-chroma artefacts. A chroma trap filter is also available. On CVBS and S-video inputs, the user can control video characteristics such as contrast, brightness, saturation, and hue via an I²C host port interface. Furthermore, luma peaking (sharpness) with programmable gain is included, as well as a patented chroma transient improvement (CTI) circuit.

It also generates synchronization, blanking, field, active video window, horizontal and vertical syncs, clock, genlock (for downstream video encoder synchronization), host CPU interrupt and programmable logic I/O signals, in addition to digital video outputs, and includes methods for advanced Vertical Blanking Interval (VBI) data retrieval. The VBI data processor (VDP) slices, parses, and performs error checking on teletext, Closed Caption (CC), and other VBI data. A built-in FIFO stores up to 11 lines of teletext data, and with proper host port synchronization, full-screen teletext retrieval is possible.

This versatile and near-professional device is applied in Digital TV, LCD TV/monitors, DVD

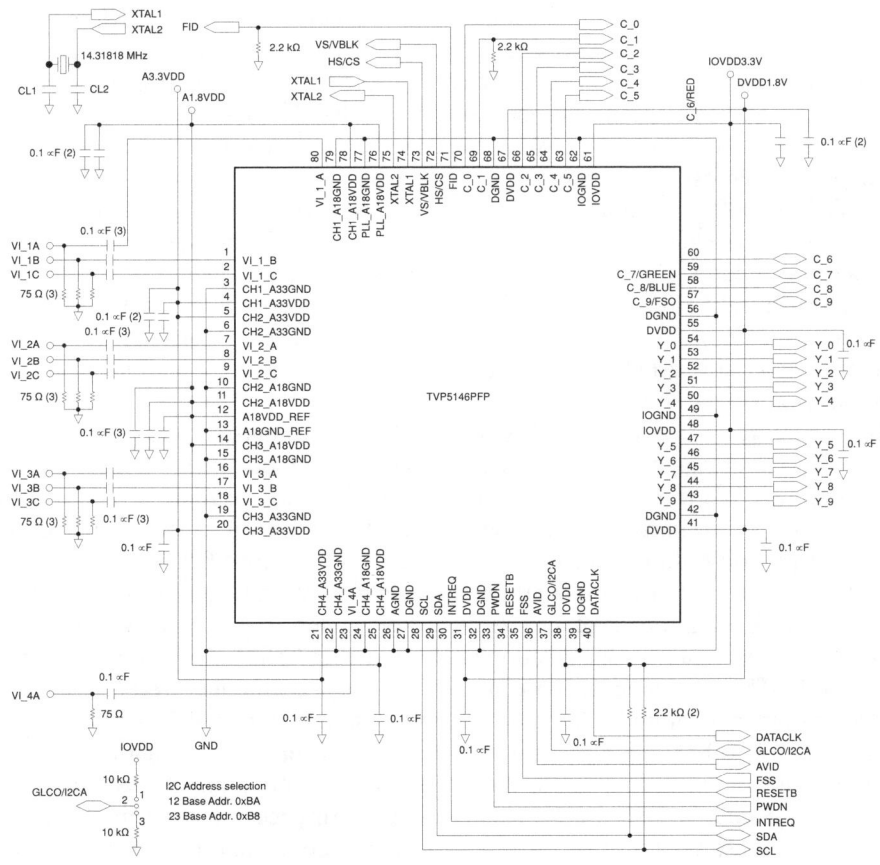

Fig. 48

NOTE: If XTAL1 is connected to clock source, input voltage high must be 1.8 V.
Terminals 69 and 71 must be connected to ground through pulldown resistors.

Players as well as PC video cards, video capture/editing and video conferencing. A typical application note schematic is shown in Fig. 48.

These all-in-one ICs are intended to be used in economical, domestic or semi-professional multisystem video equipments with modest needs and they can be easily found on market as television spare parts. Professional and broadcast TV monitors continued to make use of separated colour decoding stages with very sophisticated and meticulous alignments.

3.10.4 Digivision TV chipset

In 1980 the engineers of the German *ITT International* began the development of specific fully digital ultimate integrated circuits after four years of conceptual development and testing of simulation, and the first samples were tested successfully in early 1981.

The high integration of these chips, codenamed *DIGIT2000*, allowed condensing in an area of about 180 cm^2 some tens of thousands of semiconductors.

Then a new TV circuitry was born using the DIGIT2000 chipset and took the name of *Digivision*.

Notably ITT from the early '80s made a big advertising promotion of the new TV sets based on this futuristic technology. Many thousands of Digivision televisions were sold over the years with the Digivision badge on the bottom left of the fascia below the CRT screen. With the registered name Digivision, also marketed under the brand names of *Graetz*, *Ingelen* and *Nokia* (adopted for some time by the French *Thomson* in some chassis equipped onto a range of high cost television sets too), ITT had produced the first digital signal processing television, which has long been talked about as a revolutionary stage in the evolution of the colour TV receiver.

It happened almost 30 years ago!!

The DIGIT2000 consisted of seven integrated circuits with a high degree of integration, and a minimum of additional components, which together made the complete processing of the signals in the video, audio, deflection and power stages. It included the following chips: the *MAA2000* (Central Control Unit, CCU), *MAA2100* (Video Codec Unit, VCU), *MAA2200* (Video Processor Unit, VPU), *MAA2300* (Audio Analogue/Digital Converter, ADC), *MAA2400* (Audio Processor Unit, APU), *MAA2500* (Deflection Processing Unit, DPU) and *MAA2600* (Clock Generator) (Fig. 49).

The advantages of the Digivision system could be summarized in the following headings: digital processing of signals; use of high degrees of integration (VLSI); limited number of external components to the system; no more critical calibrations due to component tolerances; absence of changes in characteristics; programmability; all software-based regulation with more comfort of use; computerized technique for the development of digital components by the manufacturer; stereo or bi-channel audio; adaptable to all standards (PAL, SECAM, NTSC); flicker-free images via a means of intermediate storage and subsequent reproduction at higher value line frequencies; automatic deletion of ghost images (reflections); improved image quality in a standard NTSC receiving by using an internal digital comb filter; reduced and contemporary vision of other program in the same context (*Picture In Picture*, PIP); freeze frame stored in memory; zoom; 1,000-line images; automatic VCR programming; digital processing of audio signals; reproduction of captions onto the screen such as videotext, teletext, etc.

All these characteristics confer outstanding performance not least the ability to lock in frequency and phase signals of very low quality and intensity, and to control screen geometry parameters acting directly by remote control, or to obtain other special PIP features, etc. Such

3.10 Colour decoders chips

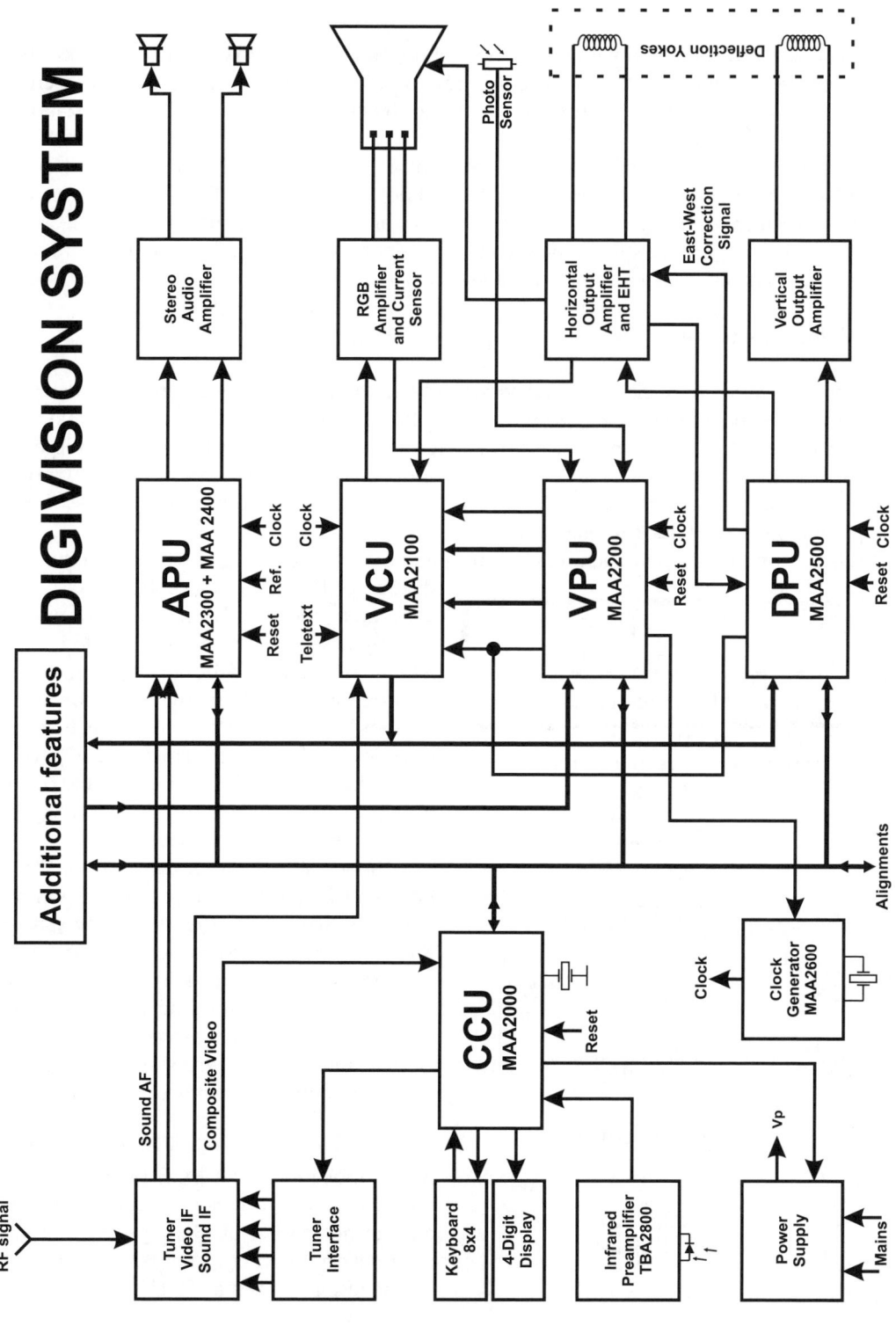

Fig. 49

Chapter 3. Colour Standard Systems

capabilities are now obtained with much more abridged and economic technologies and those ones proposed by the ITT at that time are now obsolete! However it represented a crucial milestone that had offered electronic engineering the possibility of a concrete realization of digital televisions.

A considerable volume of Digivision televisions are still in homes today and fully working. Nowadays other manufacturers promote new technologies based on digital electronics. Some citizens, particularly the older and less technically aware may believe they are today ready for digital switchover as a result of owning such products not knowing that these last ones are based upon a '30-years-old legacy digital technology' as Digivision!

3.11 Video conversion

From the earlier fully electronic television system, the composite video was inevitably the leading video signal used only by the professional broadcasters and later even by simple users, when with the reduction of costs it was possible for the general public to buy video equipments.

Later S-Video standard was introduced and some attempts have been made to support it. However initially it has been primarily limited to high-end, professional and broadcast S-VHS equipments such as VCRs, cameras and TVs.

A large numbers of manufacturers began to support S-Video standard when, with the introduction of DVD players, capture & displaying computer video cards, CCD-based cameras and digital set top boxes, most detailed images were generated. As a bonus, a common user could choose an alternative of vision in the form of YUV or RGB output signals that such new devices furnished, eliminating in a single stroke the NTSC or PAL encoding and decoding flaws and artefacts, with even a sharper and clearer vision than S-Video.

The so-called *Peritel* or *Euroconnector*, better known as 21-pin SCART connector, was proposed by *Syndicat des Constructeurs d'Appareils Radiorécepteurs et Téléviseurs* and introduced after 1978 for the PAL & SECAM consumer video market for grouping altogether both CVBS and RGB inputs as well as outputs (including audio stereo signals) to standardize all connection and connectors in a unique cable and socket (Fig. 50).

In the case of using a display device in RGB input mode, the RGB signals for feeding such input are generally at 0.7V pp level and do not have a blanking pedestal or sync information. In order to render stable pictures on the display screen a fourth signal had to be

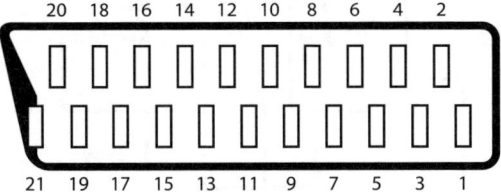

FEMALE CONNECTOR SEEN FROM THE FRONT

PIN OUT

Pin 1	Audio output (right)
Pin 2	Audio input (right)
Pin 3	Audio output (left/mono)
Pin 4	Audio ground
Pin 5	RGB Blue ground (pin 7 ground)
Pin 6	Audio input (left/mono)
Pin 7	RGB Blue up
	S-Video C down
	Component Pb up
Pin 8	Status & Aspect Ratio up
	* 0–0.4V → off
	* 5–8V → 16:9
	* 9.5–12V → on/4:3
Pin 9	RGB Green ground (pin 11 ground)
Pin 10	Clock / Data 2
	Control bus (AV.link)
Pin 11	RGB Green up
	Component Y up
Pin 12	Reserved / Data 1
Pin 13	RGB Red ground (pin 15 ground)
Pin 14	pin 12 & pin 16 ground
Pin 15	RGB Red up
	S-Video C up
	Component Pr up
Pin 16	Blanking signal up
	RGB-selection voltage up
	* 0–0.4V → composite
	* 1–3V → RGB
Pin 17	Composite video ground (pin 19 & 20 ground)
Pin 18	Blanking signal ground (pin 16 ground)
Pin 19	Composite video output
	S-Video Y output
Pin 20	Composite video input
	S-Video Y input
Pin 21	pin 8 & pin 10 ground

Fig. 50

3.11 Video conversion

utilized for conveying composite synchronism which will coincide with the CVBS signal.

In the case of NTSC high-end consumer video equipments, analogue CVBS, S-Video, YUV and RGB video signals are furnished on separate connectors, however. There is also a difference in RGB mode, i.e. the presence of a small blanking pedestal on each colour channel and optionally an additional composite synchronism included in the green video signal.

Recently also low-end and low-cost DVD players are being equipped with separate outputs for the European market in addition to the traditional SCART connector in order to show television images onto several low and high end display devices. However there is some evident confusion about labelling the colour difference signal: YUV, YR-YB-Y, YC_bC_r or YP_bP_r!

In some previous paragraphs, we saw how to get RGB signals from CVBS/S-Video (decoder) and vice versa (encoder). Below are explained some signal 'conversions' methods between 'intermediate' video signals, i.e. from S-Video to CVBS or from YUV to RGB and vice versa.

3.11.1 From PAL S-Video to CVBS and vice versa

The simplest way to create a PAL composite video from a PAL S-Video source is shown in Fig. 51, yet this system owns a serious drawback. In particular, wherever the image has large fine detailed zones like a T-shirt with fine stripes or a check pattern, then we will see a very noticeable yellow/purple moiré interference pattern. This visible interference is the *cross colour* artefact caused, as explained previously, by the beating between the higher frequencies in the luminance signal, which for a S-Video source can rise up to 5MHz or more, and the 4.43 MHz frequency subcarrier in chrominance signal.

To avoid this, the best approach is to reduce cross colour interference using a simple LC trap circuit centred on the 4.43MHz colour subcarrier frequency to cut off all the frequencies near and around the luminance signal, removing most of the higher luminance frequencies that cause cross colour patterning, while leaving the luminance frequencies below about 3.8MHz and above 5.5MHz untouched. In Fig. 52 is an application based mainly on Maxim's dual video operational amplifier U1 (MAX4451ESA) available from Futurlec (www.futurlec.com). U1 could be substituted by another low cost amplifier with same features.

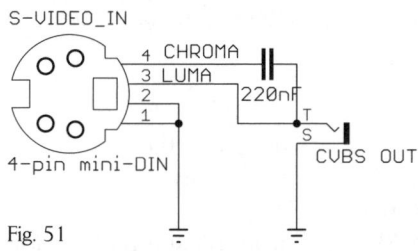

Fig. 51

Every op-amp is 'flybacked' via R3/R4 & R5/R6 to get a x2 gain to compensate for the loss due to the 75 Ω input loads. When C2 is adjusted to resonate at 4.43361875MHz with L1, this LC circuit forms a low-impedance path to ground at that frequency. The best way to achieve this is to input a CVBS source to pin 3 of the mini-DIN S-Video IN connector and at RCA_CONN output connect an oscilloscope's probe and route C1 until you see on the instrument's screen the subcarrier signal level drop down into a null value. Another less refined method is to visually judge on a CVBS TV monitor a video image affected by cross-colour interferences (coming for example by a DVD still frame) and calibrate C1 until the cross-colour artefacts disappear. Note that this circuit can be easily adapted for NTSC by changing L1 value to trap the 3.58 MHz subcarrier. U1 needs a regulated dual voltage of ±5VDC respect to ground and a lot of attention to solder it because it is a small SMD device.

Chapter 3. Colour Standard Systems

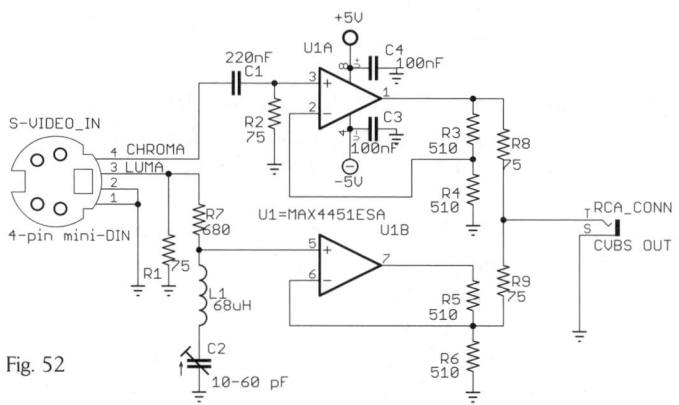

Fig. 52

The inverse operation, i.e. from CVBS to S-Video or Y/C, was explained previously speaking about RGB decoding technique in §10 (Fig. 36 - RGB Decoder - Input stage 1/3) by using simple RCL filters. However the comb filter, as we have seen, is the best solution to separate Y & C from CVBS and we suggested in Fig. 39 to use the TDA9181 for this. Some improvements beyond TDA9181 comb filter have been performed in the years, because by using this simple comb filter, even if it increased horizontal resolution, it hung dots at colour boundaries or vertical dots crawled with severe artefacts at vertical colour transitions.

New devices were invented called *2-Line* (or *2D*), *3-Line* (or *3D*), *2D Adaptive* and *3D Motion Adaptive* comb filter ICs. The term '2D' indicates that the filter implements detection of both horizontal transitions (along horizontal lines) and vertical transitions (between horizontal lines) within the displayed pictures. 3D motion adaptive comb filters represent the most sophisticated device available being based on digital technology. They are surely more expensive if compared to all previous filter types but they can provide a quasi-perfect separation of Y and C for still images ensuring also best horizontal and vertical bandwidth without artefacts, even if the latter are still possible around moving objects.

We suggest consulting the respective website application notes of the following ICs for more hints: *ML87V21071* (OKI Semiconductors), *TW9919* (Techwell), *ADV7802* (Analog Device), *SAA7114/SAA7115/SAA7118* (Philips Semiconductors).

3.11.2 From YUV to RGB and vice versa

The YUV to RGB conversion and vice versa is virtually lossless. This is why the conversions are obtained by simple precision matrix calculus by means of some resistor network and buffer amplifiers. An example of YUV to RGB conversion is shown in Fig. 53.

The circuit is based on the four dual video amplifiers MAX457 from Maxim packaged as DIL8 that can be substituted by the MAX4451 SMD version, or any other equivalent video compliant device. IC4A derives via a resistor network the G-Y signal and IC1A, IC2A & IC3A create -R, -G & -B signals while IC1B, IC2B and IC3B amplify and invert the polarity of RGB signals to drive a 75 Ω load. T1 and T2 extract composite synchronism from Y to feed IC4B used as a sync buffer. Except for R34, R35, R36, R37 and R38, all resistors used MUST be 1% or less of tolerance. A dual well filtered and regulated ±5VDC voltage power supply is also needed.

Recently National Semiconductors introduced the LMH1251, a YUV to RGB Decoder and 2:1 Video Switch IC. The LMH1251 is a high-speed triple 2:1 video multiplexer with an integrated sync processor and colour space converter (Fig. 54). One input channel accepts standard RGB+HV (separated syncs) PC graphics video and the second input channel accepts YUV component video. If the first input of the MUX is selected, the device will output the RGBHV video from the input with unity gain. If the second input of the MUX is selected, sync processing and colour space conversion will be performed on the YUV component signals to provide an equivalent RGBHV signal at the output. The device is packaged in the very small 24-pin

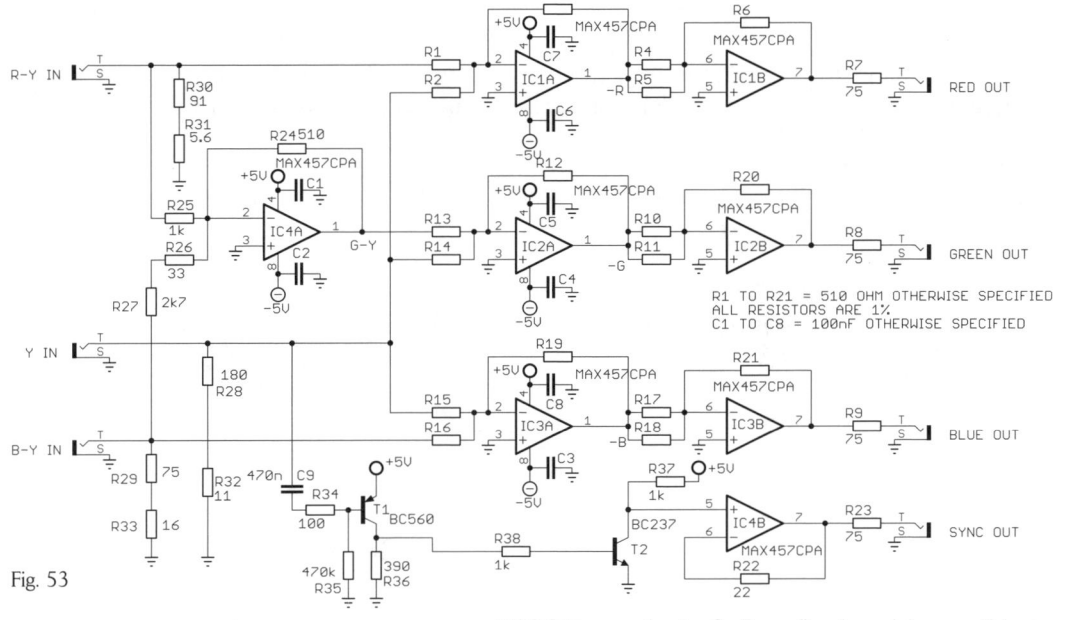

Fig. 53

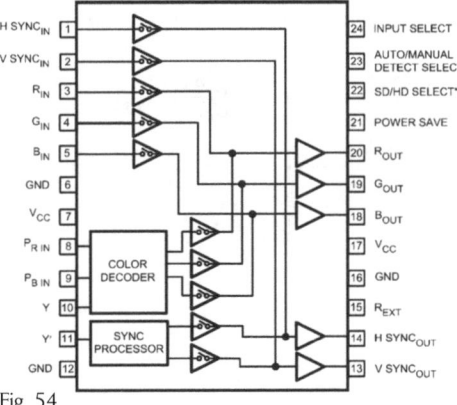

Fig. 54

TSSOP case (only 8x5mm!), almost impossible to solder by hand. National may send some samples on demand to experiment with this device. Of course, most application notes are to be found on National's website.

The inverse operation, i.e. from RGB to YUV is simpler. In Fig. 55 the RGB to YUV circuit is based again on two MAX457 (or any other equivalent chip) and the green signal must include the composite syncs. Following all the previous hints given for YUV to RGB Converter, we believe that this circuit does not need more details here for clarifying and building it.

3.12 Some final considerations

At the beginning of this chapter we explained the various methods developed by some companies to merge colour information with the existing well-established greyscale broadcasts, forming a single encoded composite signal ready to be transmitted, recorded or processed otherwise. With the digital age, the actual trend is to keep the three RGB/YUV component signals separated before its digital (satellite or terrestrial) transmission while the receiving device (set top box) reconstructs the RGB/YUV signals to feed analogue RGB/YUV-fitted equipments such as a plasma TV set or domestic video projector.

However the TV broadcast services still tend towards the use of analogue component signals in the chain of production, post-production or live transmission relegating to digital media the footage storing and using a digital MPEG encoder the 'simultaneous' broadcasting of all analogue RGB signals. After many efforts, battles and trials done by genius including Baird, Goldmark, Sarnoff, Bruch, De France and companies like CBS, RCA, Telefunken and others, we are returning, with digital technology, to the initial idea to transmit the RGB signals derived

Chapter 3. Colour Standard Systems

from, for example, three pickup tubes, but without lowering the visual quality through subcarrier quadrature modulators, filters, delay lines, matrix circuits, etc., that the NTSC/PAL/SECAM colour standards have adopted.

At least when the full migration from broadcast analogue devices (expensive and still fully working!) to digital professional equipments will be possible to achieve at a reasonable price!

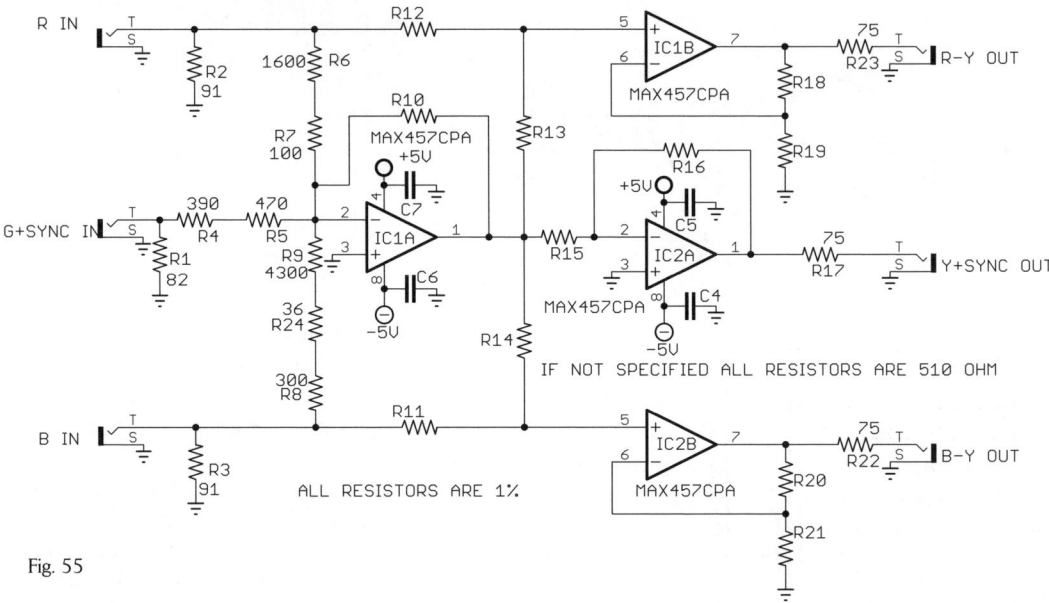

Fig. 55

4. Pattern & Monoscope Generators

Since television was first introduced in 1929 with the first experimental service broadcast by the *BBC* in collaboration with *The Baird Company* (thus based upon Baird's mechanical television prototypes), the need for *Test Patterns* (called also *Test Cards*) to examine the transmission quality or calibrate the apparatus was evident, appearing for the first time around 1934. One of these consisted of a chart with a simple circle above a horizontal line, which tested the picture aspect ratio, and another of the first was a triangular frequency grating for judging high-frequency response. We should bear in mind that it was the time of the first 'portrait' television, since the standard picture ratio was very nearly at a 7:3 aspect ratio, i.e. the image was much taller rather than wider.

In 1937 BBC abandoned the Baird's mechanical system in favour of the 405-lines *Marconi-EMI* 'high definition' and fully electronic system. It transmitted for the first time a 405-lines *BBC Tuning Signal*, which was used to identify the station and provide something of interest for viewers waiting for the equipment to 'warm up'. In the following years BBC started regular program broadcasting using several variants of test patterns only in the early morning and late evening hours of specific days.

These patterns were usually hand-drawn on large pieces of paper card, typically 60x90 centimetres, and placed in front of a large camera allocated just for this purpose. The signal thus generated was displayed directly on a very highly fine-tuned preview TV screen (used as terms of comparison) to allow their engineers to determine the broadcast bandwidth quality and TV repairers the receiver alignment, i.e. its linearity and convergence stages, along with other parameters. Some stations still transmit such screening signals during the first hours of the morning, but due to the transmission scheduling it is awkward for a repairer to use it in such periods, so some industries later designed and marketed 'pseudo-digital' test pattern generators provided with a high accuracy performance as independent equipment which however was (and still remains) quite expensive. The paragraph 5 contains much more informations about this.

In the previous chapters, we saw how to generate composite synchronisms (by building the SPG625) and later a basic black-raster colour composite video commonly called *Black & Burst* (by using for example the encoder MC1377). In this chapter we will examine how to add B&W and coloured simple test patterns to the 'visible' part of television lines by suggesting some elementary circuits.

One means to achieve this would be to duly program a PIC microcontroller device then able to generate everything, i.e. from CCIR625 compliant composite video to various test patterns such as grid or chessboard.

However, we will instead choose not to 'shorten the way' by using a PIC since the purpose of this book is to guide the technical reader into the construction of some circuits to understand the video signal in its numerous aspects.

A complete test pattern generator would create a plurality of figures respectively representing critical or strategic points to get the optimum response from video equipment. The amplitudes of the various signals generated are such that when the signal is applied to the input of a device, the test pattern seen on the screen will show an optimal verification check if everything has been properly aligned. By selecting frequencies properly locked to HS, the pattern also provides a good visual indication of image details. Moreover, the patterns are repeti-

Chapter 4. Pattern & Monoscope Generators

tive whereby the alignment, grey scale range and sweep linearity can be determined by simply observing the test pattern on the screen of the receiver. Preferably, the test signal also includes a relatively low video frequency band to establish the response on large areas of the screen as visual reference.

Now, how can we create some high-stability well-structured B&W and colour test patterns without using expressly programmed chips or taking a test board connected to a specialized pickup camera electronic tube?

The answer is revealed below.

4.1 The colour test pattern bars generator

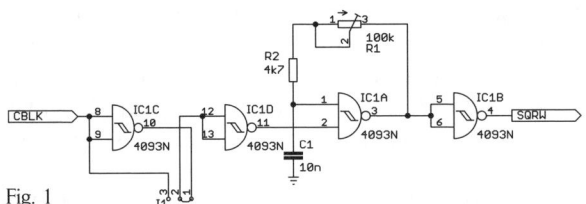

In one of the previous chapters, we have seen how to generate a greyscale pattern using only a single CMOS IC, the 4060, and later a more complex pattern generator based upon the expensive and discontinued ZNA234.

Fig. 1

However, there are many other strategies to get the same results using ordinary ICs. The simplest method to create a repetitive pattern, as vertical bars alternately black and white, is shown in Fig. 1. This simple circuit is a start-stop oscillator driven by composite blanking. One of four dual input NAND Schmitt triggers included in IC1A (4093) is left free to oscillate with a frequency established by R1/R2/C1 values when the negative-going composite blanking signal present at pin 2 moves to a logic high level. In this way a HS-locked square wave is generated at pin 3, inverted by IC1B and available at output SQRW at a CMOS level. Simply by adjusting R1, you can vary the number of vertical bars. The CBLK signal can be taken from the SPG625's output of the same name without modification but sharing the same +5VDC power supply and by connecting pin 1 & 2 of J1 together. Whereas if a positive-going CBLK signal is available, you should select pin 2 & 3 of J1 to reverse the signal polarity.

In order to view the vertical bars we need an encoder. For design simplicity the AD724 was chosen for this purpose and, by connecting together the AD724's RGB inputs and then the SQRW signal (properly attenuated by the resistor divider R3/R4) and using negative-going composite syncs, a composite video with black & white vertical bars included will be available

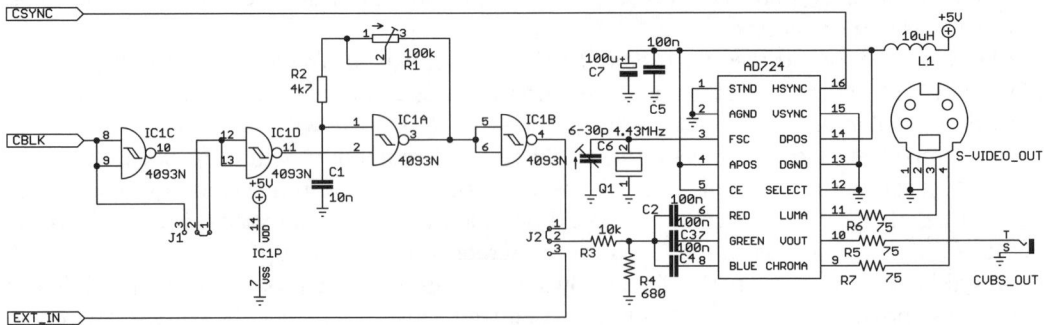

Fig. 2

4.1 The colour test pattern bars generator

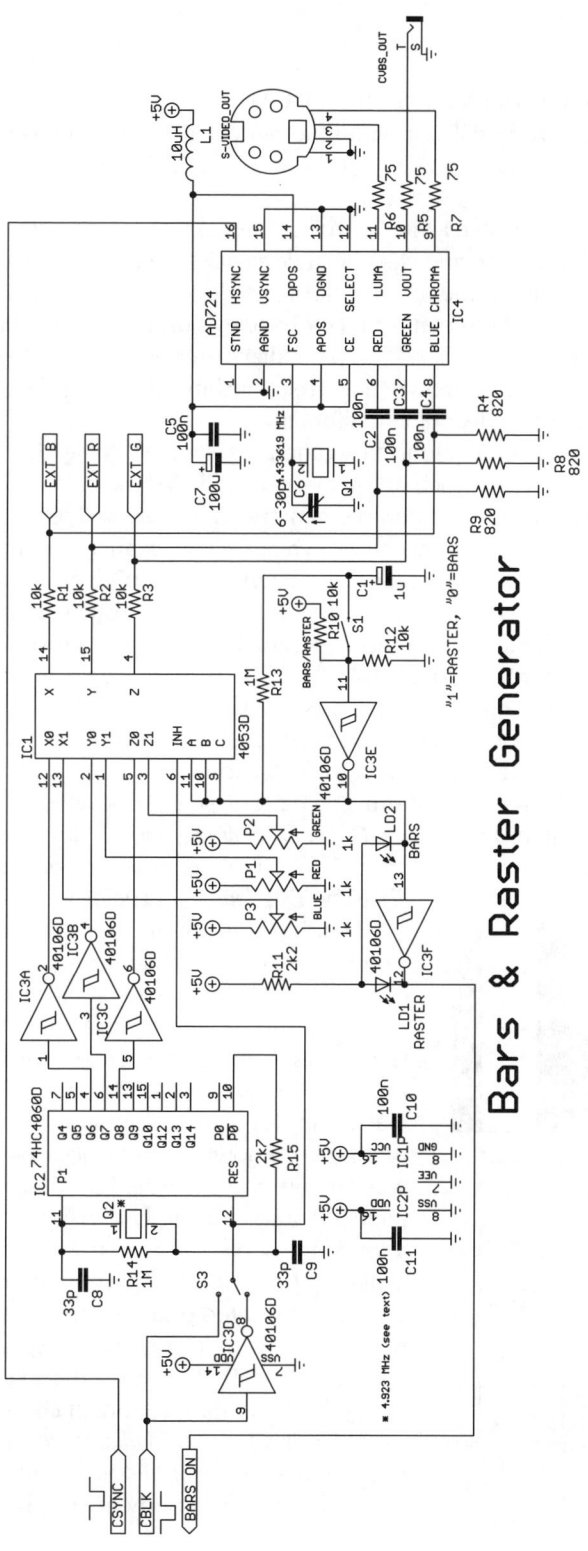

Fig. 3

Chapter 4. Pattern & Monoscope Generators

at CVBS_OUT and a Y/C output at S-VIDEO_OUT (Fig. 2).

Due to the use of a RC oscillator with inherent poor stability, the vertical parallel bars can weaver slightly. If we wish for a steady pattern, we should change strategy. A more useful and professional derivative would show a series of rainbow-coloured vertical bars (accomplished by fixed or variable coloured rasters) thus useful in the field of colour televisions where it would provide a number of luminance and chrominance signals, useful for the adjustment and troubleshooting of TV equipment. Generally, a colour bar generator produces a test signal resulting in a colour bar pattern on the screen of a colour television receiving device. Since the correct colour of each bar is known, the loss of a particular component can be observed and the test signal, having components with known phase relationships, can be easily traced through specific instrumentation to identify a decoding malfunction.

A complete schematic of a colour bars generator is visible in Fig. 3.

This circuit generates a sequence of eight equal width vertical bars from left to right showing saturated primary colours and their complementary colours as well as black and white. We will see the bars on the CRT screen in this order: white, yellow, cyan, green, magenta, red, blue and black. The standard order of presentation has been chosen to give a descending sequence of luminance values onto an oscilloscope's screen and a characteristic pattern of 'dot landings' on a vectorscope's screen. Some professional generators also provide a reverse sequence as well as a continuous 'step-less' rainbow colour figure.

This particular colour pattern was proposed by *EBU* (*European Broadcasting Union*, a confederation of 75 broadcasting organisations from 56 countries and 43 associate broadcasters) as a test pattern for balancing the colours in early PAL cameras.

However historically the first colour bar pattern was proposed in 1951 and patented in 1956 for the NTSC system by the late *David D. Holmes*, a RCA director of the *David Sarnoff Research Center* in Princeton (New Jersey, USA). Holmes was also director of the *RCA Color Television Development Division* for many years.

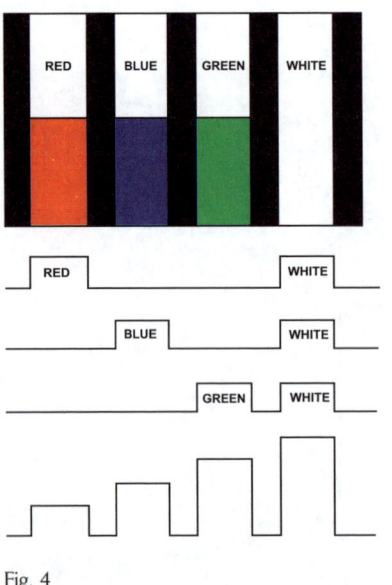

Fig. 4

Holding the patent along with *David Larky*, he invented colour bars while working for RCA, an invention curiously inspired by the spinnaker of his sailboat. At that time in the RCA research labs the scientists were using test signals from scanned slides which he regarded as suboptimal, full of noise and other rubbish, so he built an electronic test signal generator, based on fifty electronic tubes, now known as the *Colour Bar Generator*. To him it was an easy job since he had designed and built a complete TV studio at the University of Nebraska (USA) and a lot of the sub-systems in the colour bar generator were similar to parts of that. This equipment was a great hit and everybody wanted one; a company took out a license and manufactured it. The pattern displayed yet was very elementary (Fig. 4); afterwards the colour bar pattern shown in Fig. 5 was employed, originally conceived in the 1970 by *Al Goldberg* of *CBS Laboratories*, and previously categorized by the *Society of Motion Picture and Television Engineers* (SMPTE); the development of this new test pattern was

4.1 The colour test pattern bars generator

Fig. 5

awarded an *Engineering Emmy* in 2001-2002.

The EBU colour bars pulse generator was inspired by a SMPTE test pattern and produced a simple sequence of time-coincident square wave signal pulses of equal amplitude and equal bandwidth. By a suitable overlap of the pulses in certain portions of the raster and non-overlap in others, it produced the three saturated primary colours as well as the three saturated complementary colours. The timings of the pseudo-digital signals for creating the RGB bars are computed from a series of simple arithmetic operations.

The eight bars have to be shown inside the 52 μs active picture line and dividing this timing by 8 we get 52/8=6.5 μS per bar. The bar pulses have to be obtained from a blanking-locked start-stop master clock (F^{BMC}) by successive binary divisions; for blue a square waveform signal whose frequency F^{BAR} will be an integer submultiple of F^{BMC}; for red a successive division by 2 and for green by 4. A complete cycle of F^{BAR} is 6.5*2=13 μs and then F^{BAR}=1/13μs=76.923 kHz which is the clock pulse available from pin Q6 of the 14-stage crystal oscillator/divider IC2 (74HC4060, Fig.3) fed by the Q2 crystal quartz of a frequency chosen as a suitable multiple of F^{BAR}, in this case 64 and therefore F^{BMC}=64*76923=4.923MHz. If that crystal value is not readily available on the market, the nearest value is still probably useable.

We should employ a more common 5 MHz quartz crystal if instead we want to use the CBLK signal coming from the SPG625, because its CBLK timing is 12.8 μs which leaves a 51.2 μs active picture line. Doing the necessary calculations, every bar will be of width 51.2/8=6.4 μs, F^{BAR} will be 1/(6.4*2)=78.125 KHz and then Q2 frequency will be just 64*F^{BAR}=5 MHz. However, in creating the bars pulses, with a little more expedience we may utilize the SPG625's 10 MHz master clock by feeding a 12-stage ripple counter like the 74HC4040 in place of the 74HC4060 as shown in Fig. 6.

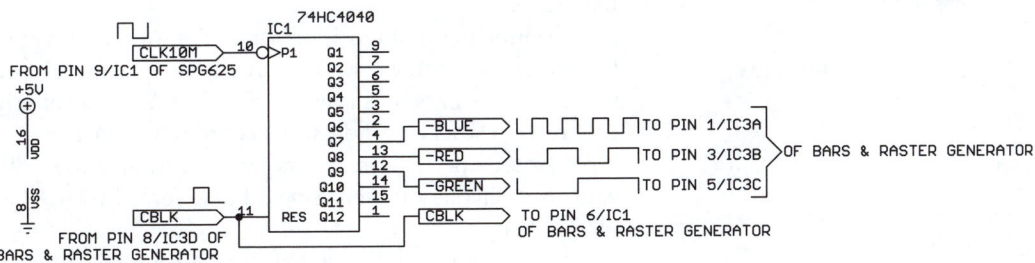

Fig. 6

The three flat-top bar pulses coming from pin 4, 6 and 14 of IC2 (or pin 4, 13 and 12 of IC1/74HC4040, Fig. 6) are reversed by the three Schmitt trigger inverters IC3A, IC3B & IC3C (40106) to feed the X0, Y0 & Z0 of IC1 (4053), respectively. The square wave RGB signals, having to pass through IC1, a triple 2-channel analogue multiplexer/demultiplexer, can be ex-

Chapter 4. Pattern & Monoscope Generators

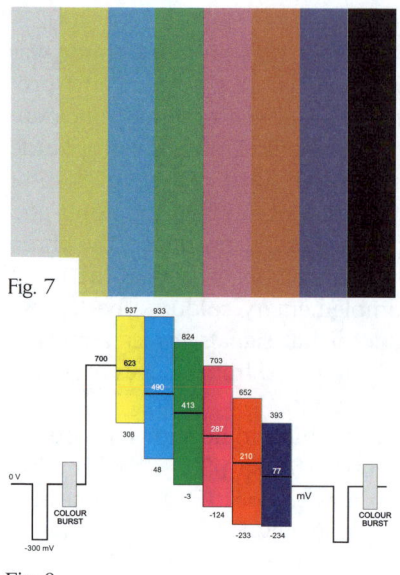

Fig. 7

Fig. 8

cluded from a logical signal '1' at the inputs A, B & C of IC1. In this way, by connecting three sources of independent variable voltages from 0 to 5V into the inputs X1, Y1 & Z1 of IC1, you can get the basic RGB colour with individually adjustable saturation via the potentiometers P1, P2 & P3 or, by changing again the logical state at pins A, B & C, the coloured bars.

To achieve this, a simple electronic on-off switch is needed by using another Schmitt trigger inverter, IC3E. By successive de-presses on S1, the logical state at IC3E's output changes because its input pin 11 is kept at half supply voltage by the divider R10/R12 made up of two equal value resistors. According to the logical level at pin 10 and through R13, the capacitor C1 will be charged or discharged. Pressing S1, causes C1 to be connected momentarily to input pin 11 thus changing the logical level according to the voltage (0 or 5V) present on C1 terminal, inverting the logical level at output pin 10 and charging (or discharging) again C1. In this way every subsequent press on S1 will produce a reverse effect, cyclically at output pin 10 of IC3E, switching then between coloured bars and raster, a condition indicated by the LEDs LD1 & LD2.

Obviously if you want to use three digital potentiometers such as Intersil's X9313 in the place of P1, P2 & P3, simply connect all their H pins to +5V and all the L pins to ground for references. The wipers will be connected singly to X1, Y1 & Z1 of IC1 for generating any coloured raster. All the others hints about to control the digitally-controlled potentiometers via push-buttons have been already furnished in the previous chapter.

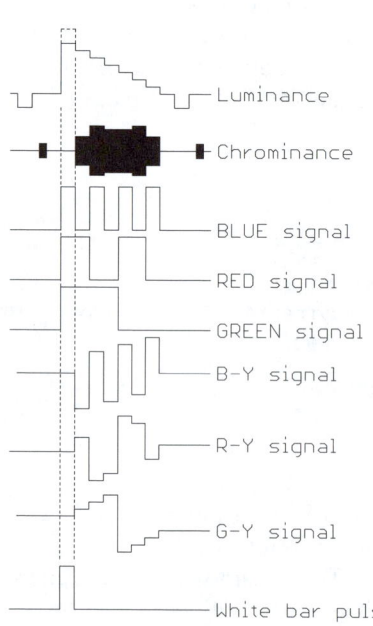

Fig. 9

The signal CBLK connected to INH of IC1 ensures that during the blanking signal no other signal is present to drive the inputs R, G and B of the PAL encoder IC4 (AD724).

Connecting a video monitor at the AD724's video out, onto the display device screen we will finally see the coloured bars as shown in Fig. 7 if C6 is properly calibrated (otherwise adjust it accordingly). In Fig. 8 is represented the coloured bars' waveform and the voltages in millivolts of every single bar referred to the 0V black level.

Albeit this circuit has been proposed as experimental, it really can be used as a diagnostic instrument for setup, alignment, calibrating, checking and repairing of PAL-only encoding and decoding equipments. The colour bars, by creating some well-distinguished luminance, chrominance, difference colour components (R-Y, B-Y and G-Y available in decoding phase) and R, G & B signals (Fig. 9),

4.1 The colour test pattern bars generator

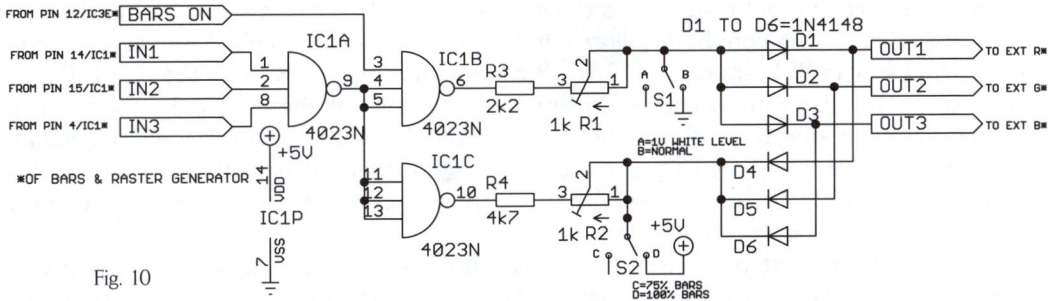

Fig. 10

are easily recognizable on the oscilloscope's screen in probing video circuitry. In addition, this series of coloured bars can be used as a reference for brightness, contrast, colour intensity, and correct colour balance of both analogue and some digital TV sets.

Generally speaking, also in association with an overlaid caption identifier, the colour bar signals can be used for testing of transmission channels too, but rather than for this precise purpose, they can be used as a means to verify that every PAL circuit of the video chain is not suffering from excessive defects in performance. Other typical uses are in camera encoders, analogue VTR output signal gain adjustments, setup, saturation (and hue for NTSC), as well as TV studio colour monitors or home receiver colour rendering.

In our *Colour Bars Generator* the video signal has 100 percent of chroma saturation; the references of 0V for black level and 700 mV for white peak along with -300 mV sync amplitude are available too. For some applications, it may be convenient to reduce the chroma saturation to 75 percent of its maximum value while in other certain cases the white bar should be elevated up to 1 V above black level. To implement these modifications in our circuit we need the add-on shown in Fig. 10.

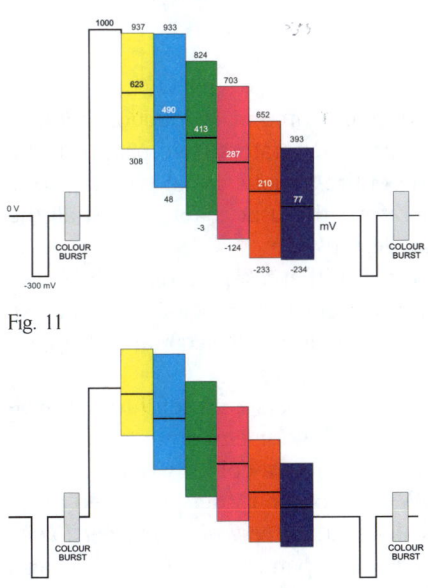

Fig. 11

Fig. 12

The add-on comprises only a triple 3-input NAND gate (4023), two resistive trimmers, resistors, six diodes and two on-off switches. The first NAND gate (IC1A) is fed by the three RGB square pulses coming from pins 14, 15 and 4 of IC1 from the Bars & Raster Generator, for which the logical state at its output changes to '0' only when the three RGB signals are initially achieve high logical level, i.e. during the white bar timing (see Fig.9). The other two NAND gates (IC1B & IC1C) are employed as buffers/inverters, which drive the trimmers R1 & R2 (together with R3 & R4) and the diodes from D1 to D6. The latter are connected so that this add-on does not interfere with the normal RGB pulses when the white bar is not displayed in all the active lines.

By a proper selection of S1 & S2, the extra gain for the white bar and the downscaled chroma amplitude are obtained. In particular if S1 is in 'A' position its amplitude can be adjusted via R1 otherwise, if 'B' is switched, the white bar pulse will be deactivated as well as 75% chroma bars are calibrated if S2 is switched in 'C' position; both calibration operations (effected empirically) need an oscilloscope. A logical level '1' at input pin 3/IC1B coming from pin 12/IC3E of the Bars & Raster Generator ensures that only when the bars mode is selected will the extra pulse for boosting the white bar be produced. Finally the three signals from the diodes will drive the PAL encoder via EXT R, EXT G and EXT B inputs of the Bars & Raster Generator. Note that IN1, IN2 & IN3 inputs or OUT1, OUT2 & OUT3 outputs are swappable between themselves. The resulting waveforms are shown in Fig. 11 and 12.

In composite or S-Video television, colour bars are even used as a sign for the leader of a recorded tape. This allows a VTR playback operator to optimize the characteristics of the signal before the program starts. In an analogue camera, you can use the colour bars to verify and optimize its encoder and they are also useful for matching the output of two or more cameras in a multi-camera mixing system. In the early and unstable receivers, colour bars have been used by consumers to adjust approximately the contrast, saturation and brightness (and hue for NTSC users) of the picture tube after a short warm-up period and before the beginning of a TV transmission. Modern receivers are sufficiently stable from first cold power-on and do not require such adjustments, except for exacting needs.

In television and professional video production, the colour bars are also used in combination with a 1000 Hz audio tone for proper levelling of the input stages of both video recording equipments and audio mixer. In the beginning of a videotape there should always be present about a minute of colour bars and tone for the correct calibration of the video player. All the professional cameras with or without embedded video recorders (the so-called *Camcorders*) are equipped with such a unit (*Bars & Tone Generator*). Sometimes the audio tone is a modulated signal that identifies the origin of the signal itself and the number of audio channels available. Two square wave audio tones of about 2400 Hz and 1200 Hz can be taken respectively from SPG625 at pins 12 & 2 of IC1 for adding this feature to our Bars & Raster Generator.

4.2 Patterns for TV set calibration

Video pattern sources are used as a reference video signal source in production lines, as well as development and TV service laboratories. Their relative generators have proven to be an ideal solution for the challenging task of providing the entire various signal types needed through their precise, repeatable images and timings. Therefore, the user is able to test all the inputs and features found both in old and modern display devices.

In addition to the above, more commonly identified problems that test pattern can help you quickly to resolve, it can help you in discovering compression artefacts (for digital equipments), linearity, RGB beam convergence and gamma (for B&W and colour TV sets), and dropped frames (for video recorders), along with many more potential problems.

Generally the colour repetitive patterns such as colour vertical bars are useful for controlling only the good quality of both the decoder and encoder of colour equipments such as a colour TV set, a television colour camera, a colour video recorder or a colour video mixer. In order to calibrate the other parameters found in both B&W and colour TV sets, we need other iterative sequences of images to be generated electronically such as *vertical & horizontal lines*, *grid* (or *crosshatch*), *dots* and *chessboard* rasters. Some professional pattern generators add also an empty circle figure positioned at the raster's centre (often coupled with smaller circles at the

4.2 Patterns for TV set calibration

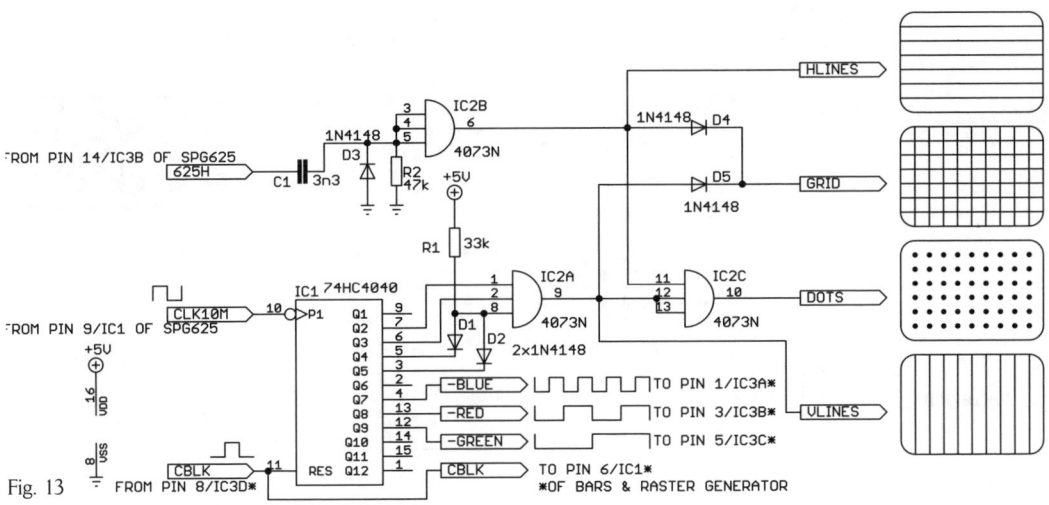

Fig. 13

four corners) and an increasing frequency multiple burst pattern.

Indeed most TV sets, besides sync extraction and colour decoding circuitry, own CRT beam deflection and extra high voltage stages which need quite precise calibration. An entire chapter will be dedicated about the argument; here we will limit ourselves to describing only how to design the patterns that permit achieving these important CRT settings: geometry, focus, correct aspect ratio and dimensions, centring, transient response and RGB beam convergence of picture. A good video technician knows how to benefit from these figure patterns and what to do in repairing a TV set.

In order to create an electronic image to be used as a reference for calibrating some CRT stages of a TV set using visual judgements, we need to design a circuit able to create an elementary image made up of 16 vertical per 12 horizontal narrow lines so to form a grid field. The numbers 16 and 12 are chosen as multiples of 4 of the 4:3 picture aspect ratio to obtain the correct proportions between the two CRT X & Y axes. For 16:9 screens obviously we need 16x9 lines or their multiples. We can get the relative signals by building the circuit shown in Fig. 13.

An additional IC (4073) and a few more components added to the circuit of Fig. 6 will furnish all the generating signals necessary to create the basic patterns and their combinations. It is based upon three 3-input AND gates which will feed at the same time the X1, Y1 and Z1 inputs of IC1 (4053, Fig. 3) thus to obtain white figures over a black background. A series of pulses at 625 Hz coming from pin 14/IC3B of SPG625 and buffered through IC2B provides the horizontal lines while an AND function made of IC2A, D1 & D2 creates from some sequential outputs of IC1 another series of 200 ns wide pulses every 3.2 µS which represents the vertical lines. D3 & D5 combines both lines to create a grid pattern whilst the remaining AND IC2C produces the dots pattern.

The chessboard pattern needs the supplementary components shown in Fig. 14.

A dual-type D flip-flop with set/reset (4013) and a quad 2-input NAND Gate, together with HLINES & GRID signals, create an alternating square wave signal synchronized by vertical and horizontal syncs.

To see all these patterns we can again employ the circuit shown in Fig. 2 by connecting them one by one, through a rotary switch 1-way/5-positions, to the EXT_IN input moving J2 to 2 & 3 pins.

Chapter 4. Pattern & Monoscope Generators

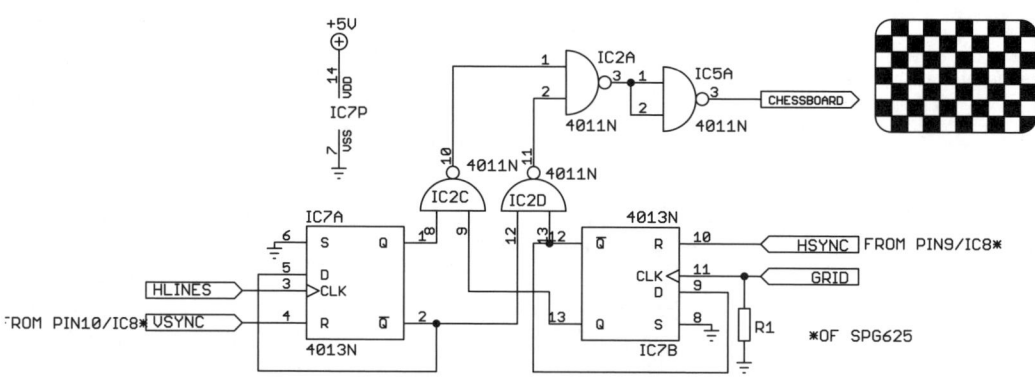

Fig. 14

4.3 Multiburst Generator

In order to test the frequency response linearity of video facilities circuits, there is a need for television broadcasting to design a unit which creates a video sweep signal to send to the input device under testing and to observe the resulting output signal in an oscilloscope's screen but this procedure cannot be used with equipment that requires the presence of sync pulses. To solve this handicap a technique widely used is to employ a multi-frequency burst signal generator (called simply *Multiburst Generator*) that produces various discrete bursts of video frequencies at full screen height or inserted into other existing patterns, as for example the colour bars, alongside composite syncs and colour burst.

The multiburst signal, which for the colour television system suitably includes a series of equal amplitude bursts of sinusoidal or square wave signals ranging from 0.5 MHz to 4.2 MHz (NTSC television system) or from 0.5 MHz to 5.8 MHz (PAL television system), is applied to the input of the circuit or system under test (Fig. 15). By observing the resulting output with an oscilloscope, the frequency response of the visual material can be determined on the basis of how the different frequency bursts are concerned or, in other words, how flat is the system's bandwidth otherwise signal amplitudes become distorted as a function of frequency, resulting in the exaggeration or attenuation of certain frequencies in the picture.

Fig. 15

It is clear that according to the type of the equipment under test, the signal has to be generated from different sources. In the case of video recorders, whether tape or disk (or even the current hard disk drive recorders) or simple amplifiers or video signal processors, the signal is generated electronically by a proper generator, which generally creates six or more packets of increasing frequencies while for DVD a test disc burned with the image samples of similar frequencies is used. For video cameras, where the signal must first be taken and then recorded, a test panel board is employed having sets of vertical B&W stripes with closer and closer spacing;

4.3 Multiburst Generator

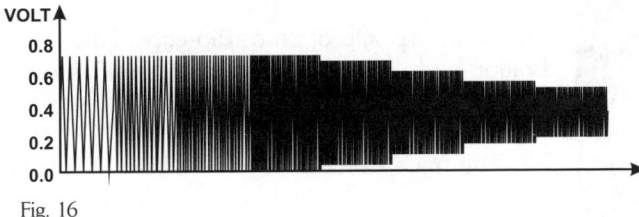

Fig. 16

by framing them, we can obtain all the desired frequencies for the various packets. The multiburst oscillogram must have the maximum amplitude for all frequencies to obtain the best result ever. A frequency of 3 MHz corresponds to 240 lines of resolution, practically the minimum condition to view an image, while a frequency of 5 MHz represents 400 lines of resolution (for example in the case of S-VHS format) which provides an excellent image in terms of resolution and detail (Fig. 16).

This multiburst signal is usually generated by successively switching a function generator or a series of oscillators on for a short time. However, the high-speed switching circuitry used produces undesirable harmonic-frequencies or sidebands that may interfere with other video equipment.

The multiburst circuit proposed in Fig. 17 produces eight groups of frequencies (square wave frequencies of 0.5, 1, 1.5, 2, 2.5, 3, 4 and 5 MHz) that are switched at intervals of 6.4 μs, producing an image in vertical stripes that occupies the entire TV screen. Pressing instead the button S1 (INSERT), the multiburst will be inserted in the centre and bottom of the image and will be about one third the screen height (Fig. 18). By switching S2, the position will be swapped while S3 reverses the polarity of the multiburst signal i.e. it starts first with a black stripe for each frequency burst or vice versa.

The main element of the multiburst circuit is the frequency-switching oscillator, which must meet the requirements of 50% duty cycle waveform with no transitory phenomenon in switching between frequencies, and a phase angle well defined at the beginning of each line, so that the TV screen can display a static image. This is particularly important because otherwise the examination of the video frequency into the equipment and its measurement would be possible

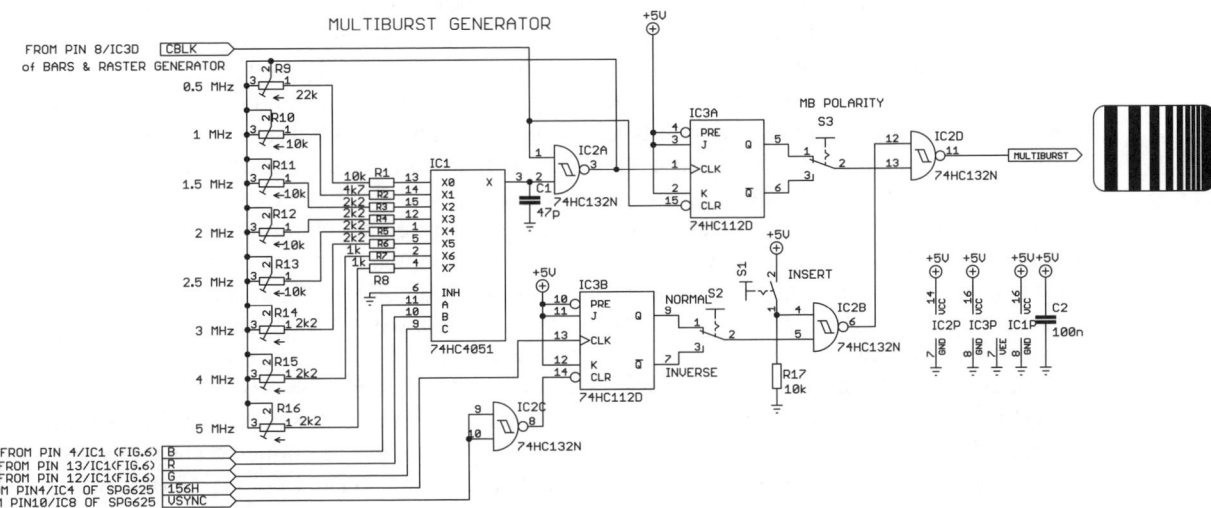

Fig. 17

Chapter 4. Pattern & Monoscope Generators

Fig. 18

only with the help of an oscilloscope. Focus calibration and visual examination of the kinescope will only be possible if the multiburst stripes are perfectly clear and static.

The main element of the oscillator is one of the four 2-input NAND Schmitt triggers in IC2 (74HC132). A HCMOS device was chosen here since in order to obtain a ratio of frequencies of 1:8 the input resistance must be very high, thus it does not load the timing circuit. The output of IC2A (pin 3) is applied to all eight resistive trimmers from R9 to R16 (along with resistors R1 to R8) connected to the eight inputs of IC1 (74HC4051), an 8-channel analogue multiplexer/demultiplexer with three address inputs (A, B & C), an active LOW enable input (INH), eight independent inputs/outputs (X0 to X7) and a common input/output (X).

The control inputs A, B & C of IC1 are driven in succession by the same signals that generate the coloured bars. Thus, the demultiplexer switches one by one in sequence the trimmer and resistors between the output of the Schmitt trigger IC2A and its input to determine the frequency of the oscillator in conjunction with capacitor C1. The correct value of each frequency is obtained by adjustment of the trimmers.

Since the waveform at IC2A's output does not have the correct duty cycle of 1:1, it is connected to one of the two J-K flip-flop included in IC3 (74HC112) configured as a divide by two. The multiburst oscillator will generate all the frequencies multiplied by 2.

The signal CBLK from SPG625 causes the oscillator to stop, while IC2A output goes to level '1' and remains in this state; thus the multiburst signal output will cease.

From pin 4 of the counter IC4A of SPG625 the signal 156H which is divided by two by the other J-K flip-flop inside IC3B, so that the multiburst will appear in the centre and bottom of the image, depending on S2's position. To ensure a perfect vertical lock, the VS pulse coming from SPG625 will reset flip-flop IC3B via the NAND IC2C (used as inverter).

This signal is logically combined in IC2D with the signal from IC2B so that if S1 is pressed then only during the period when the multiburst generator is running will the output of IC2D (pin 11) be enabled. Normally the multiburst will be always active at full screen height.

In order to calibrate the multiburst circuit you must have a frequency counter ranging from 0 to 10 MHz or more. This circuit must be calibrated on the bench, step by step, since the dynamic switching of frequencies would result in some indecipherable measurements. To do this, connect inputs A, B & C of IC1 to +5 V or ground, so to obtain all the eight possible logical combinations/permutations. You can select the various inputs X0 to X7 in sequence to connect the trimmers to the oscillator IC2A. Now adjust the relative trimmers until the frequency counter's display shows the appropriate frequency values at pin 13 of IC2D. The input CBLK must be always be kept at logic level '1' for the duration of the calibration otherwise the oscillator will remain turned off. After all the settings, connect RGB and CBLK inputs to the relative signals as described in Fig. 17 and the output MULTIBURST should be connected as previously described for the previous patterns. The result should be as shown in Fig. 15 or Fig. 18 depending on the switches' positions.

4.4 An automatic RGB colour changer

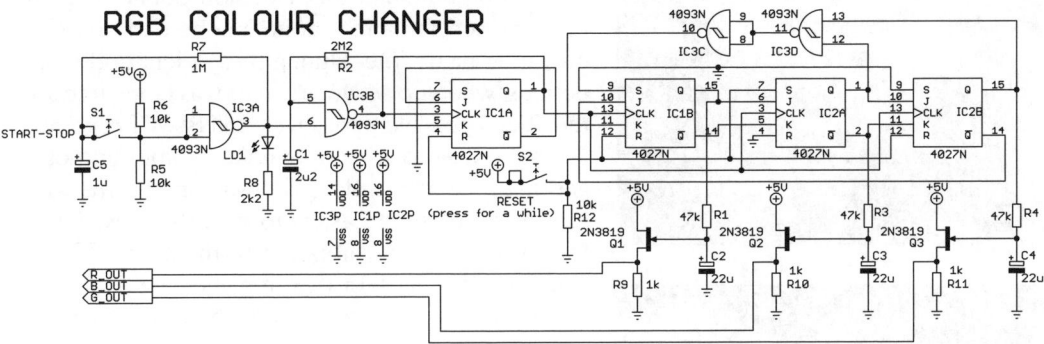

Fig. 19

In Fig. 19 is shown a curious and interesting circuit which creates a slow sequence of changing colours. A clock oscillator, made up of one of the four NAND Schmitt triggers included in IC3 (4093), running at about 0.6 Hz is divided by 2 using one of the two J-K flip-flops inside IC1 (4027) to give a 0.3Hz square wave. This drives three J-K flip-flops (IC1B & IC2) connected to provide 3-step outputs so representing red, green and blue signals. In fact the last Q output of counter IC2B is connected to the J input of the first divider (IC1B) so that a sort of pseudo-random loop counter can be created. The 3-step outputs will charge the three electrolytic capacitors C2, C3 & C4 through the three resistors R1, R2 & R4 respectively, thus generating three ramps which will provide a smoothly changing tri-phase sequence. The three FETs Q1, Q2 & Q3 as simple source follower stages furnish a 0.7 Vpp RGB video, suitable for feeding the PAL encoder. These RGB signals must include composite blanking otherwise the PAL encoder could not operate correctly.

To adapt this circuit to our Bars & Raster Generator we should simply disconnect the wipers of the three potentiometers P1, P2 & P3 and connect R_OUT, G_OUT & B_OUT of RGB Colour Changer to pin 1, 3 & 13 of IC1 of Bars & Raster Generator respectively. S1 should be switched to raster mode if we want to see the changing colour effect. The R_OUT, G_OUT & B_OUT outputs are not binding and can be exchanged between themselves to visualize other colour sequences. If needed, the speed of colour turnover can be increased or diminished by varying the value of R2. The sequence can be started by pressing S1 and frozen at a particular colour (from the eight possible) by pressing S1 once again, indicated by the LED LD1. S2 will reset all the flip-flops to an initial stage (black raster) if it is kept held for a short while so to give the capacitors time to discharge completely.

This circuit is surely the clearest proof of the greater possibilities of an analogue video compared to digital where for this particular purpose, advantage is taken only of simple increasing and decreasing voltages based on successive tiny micro steps instead of a continuous and quasi-uniform capacitor charging which would be made possible by an analogue technique.

This RGB Colour Changer can be employed as a fun background generator for an overlay titling system.

Chapter 4. Pattern & Monoscope Generators

Fig. 20

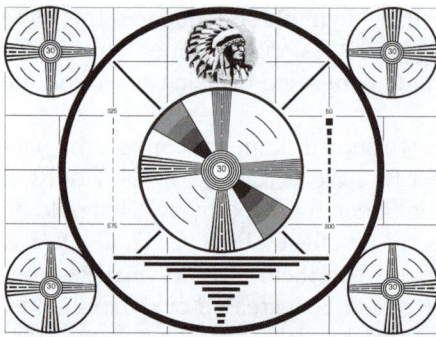

Fig. 21

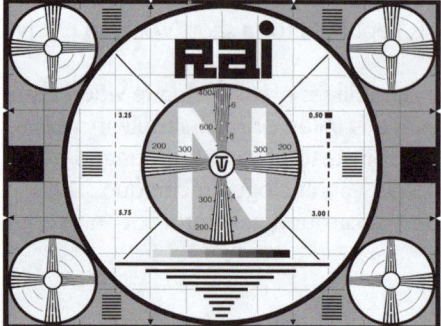

Fig. 22

4.5 Test cards and monoscopes

A long time ago the television public programming ceased for a few hours of the early morning or during night-time. During these silent and empty periods, when no active program was being transmitted, it was possible to receive a still B&W image comprising of a set of circles, bundles of lines and other geometric elements, alongside with an irksome audio tone, used by broadcasters to provide a test pattern for facilitating the calibration of the studio TV sets and by technicians for their repairs.

As previously mentioned, this still image was produced in early days by framing a test paper card with a dedicated camera but this wasteful system was abandoned when a special kinescope called a *Monoscope* was invented. Essentially, it was similar in construction to an ordinary monochrome cathode ray tube but driven backwards (working so likewise and conceptually to a pickup tube or *mono*chrome icono*scope*, from which its name was derived). It shared the property of being a picture signal makeup but in place of a photosensitive target it had a metallic surface representing the tube's anode upon which a pattern or photograph was printed or etched and which, by scanning the target via an electron beam with precision deflection, generated a video picture signal corresponding to the printed image. These were fragile, careful, elaborate and of costly construction, but had the advantages to be small, reliable and the monoscope image was always properly framed and in focus.

However these tubes could only handle black and white images, with no shades of gray, such that the halftones had to be simulated with a cluster of points or a sequence of lines progressively narrower. Depending on the scanned pattern, the black outlines stopped the flow of electrons to perform a corresponding 'pulsing' video output at the anode (Fig. 20 & 21). However it was found that the secondary emission rate of the target material varied with age and use and such variations were most noticeable over those sections of the target which were more frequently scanned. Since this degradation was not uniform over the surface of the target's outlines, the

Fig. 23

4.5 Test cards and monoscopes

end result was a decreased video output signal at a given beam current, which eventually rendered the tube unusable or, in other words, the effective tube life was shortened. Consequently, other later inventions provided improved monoscope tubes which overcame the disadvantages and deficiencies of prior art devices before the semiconductor age.

Many television stations used this type of generators when they were making their productions entirely in black and white; the most popular stations often specially ordered monoscope tubes with their initials and origin identifier (logo) inserted by the factory above or below the small circle. Even some major networks have used such patterns or slides created for their specific purposes.

Probably the first and most famous tube-based monoscope was the so-called *Indian Head Monoscope*, an American B&W monoscope pattern (Fig. 22), originated by RCA in 1939.

Curiously, many years later the Japanese borrowed the *Indian Head Monoscope* and improved on it. *ShibaSoku* (and possibly others) had designed and built a solid-state monoscope generator with higher resolution, real greyscale (instead of false gray tones made by tiny dots), markers to indicate the level of overscan and in some versions a grey background (outside the circles) instead of a white one. This model could be recognized by means of the dragon's head in the place of the Native American and the words '525 lines' or '625 lines' depending on the model.

In Italy the monoscope of the national TV channel (RAI) was another Indian Head's variant (Fig. 23); in UK the first monoscopes (here called *Test Cards*) were designed to work with the early B&W 405-lines standard (tube-based and in 5:4 aspect ratio) and later for the PAL CCIR625 system (now semiconductor-based and in 4:3 ratio), named in alphabetical order from 'A' to the latest 'J' type. The 16:9 version of the 'J' type was named with the letter 'W'. A high-definition version of Test Card 'W', called Test Card 'X', was also designed. Starting from 'F' type (the first in colour), all these Test Cards incorporate at the centre a colour photograph of then nine years old *Carole Hersee*[1] (daughter of BBC engineer *George Hersee* who designed the card in 1967). She is seen to play noughts and crosses with a doll (Fig. 24). These test cards have been used both by the BBC and ITV.

When colour broadcasting became popular, many TV stations chose to not employ more such monoscopes because it gave a feeling of 'stale stuff', and preferred instead to opt for simple coloured bars (created using a lot of thermionic valves, see previous paragraphs), since a complex colour tube-based monoscope would be too burdensome to buy in those early colour days, and even more if designed to be sold later worldwide.

Some TV stations used the traditional test patterns right until around 1970, alternating between B&W monoscopes and colour bars, often during the night transmissions along with a superimposed logo or caption identifier of the television station made using a specialized character generator. Later, the first totally solid-state colour monoscope generators were marketed in large scale amongst which the monochrome Philips *PM5440* was popular and surely the most famous was the *PM5544*. Incidentally, Philips designed a new series called *PM5644*, which generated the same scheme made famous by the PM5544 but converted into 16:9 format that some European broadcasters (such as RAI and BRT) are still using (Fig. 25). Before the days of microprocessors this was a complex and expensive 'black box' of electronic trickery! Due to the evolvement of electronics, simpler and less expensive solid-state monoscopes were designed such as the Fernseh/Telefunken *FuBK* (*Funkbetriebungskommission*, the German Television Service Commission, Fig. 26 and the *ETP-1* from *Independent Broadcasting Authority* (IBA), manufactured by some companies (Fig. 27). These pieces of equipment did not only create the monoscope as clean, clear and

Chapter 4. Pattern & Monoscope Generators

Fig. 24

Fig. 25

Fig. 26

Fig. 27

beautiful kaleidoscopic pictures to be seen on the TV screen, but they were also excellent tools that allowed diagnosis of many faults that could arise in a TV or a monitor set during its life, much more efficiently than transmitted monoscopes created using 'beam-pelted' cards embedded in a dedicated CRT which for its nature, as it was energised on 24H at day, changed performance during its relatively short life.

Furthermore, because the cost of a TV colour set was not easily affordable, it was difficult to convince a potential customer to buy a colour television by showing its blank screen in a TV shop. Even if a customer would buy it blindly, the seller would have to install it for him, erect an aerial and perform a complicated series of adjustments before receiving the first pictures. Never forget that the reliability of early TV sets was somewhat 'touch-and-go' and it was very lucky for the unit to work when first switched on! Even if it did work, there were adjustments for height, width, linearity, synchronisation, contrast, saturation, hue and focus to be made. The TV set took several minutes to warm up, and during this period the picture could vary quite a lot and even onwards the electronic stability was a mirage! So a coloured monoscope visible on a screen of a 'perfectly working' TV colour set permitted a potential purchaser to judge before buying and choose between various model and brands with the viewer participating to tweak by himself the user controls before the beginning of TV programming.

In practice all of the solid-state electronic versions of monoscopes included essentially all the simple patterns which we have previously talked about, i.e. grid, chessboard, colour bars, multiburst, etc., with the evident difference of having grouped them harmonically in a unique design and mostly comprehensible for an expert TV technician.

The intervening years have spawned many variations on the PM5544 pattern, and on satellite channels it was rare to find any two identical. Many of them are so different from the original that the monoscope seemed to be a full-screen logo TV station identifier rather than a pattern for helping the TV set calibrations!

Today the monoscope has almost disappeared from television broadcasts because TV repairers are now equipped with such tools able to create similar test patterns (due also to price fall) and send it to the equipment to be controlled through numerous available inputs (RF, SCART, Y/C, VGA, etc.); in addition modern TV sets do not require more user calibration following a 'warm-up'.

On the contrary, in most of the Third World's television networks and stations, monoscopes (or similar units) are still diffused and transmitted because they do not have 24-hour programming and the relative populations still own old TV sets.

Indeed formerly a common sight, test cards are now only rarely seen outside of television, post-production and distribution studios. In particular, they are no longer intended to assist normal viewers in the calibration of television sets.

However monoscopes, colour bars and certain test patterns are still commonly used within television production facilities. Many TV networks worldwide still have an analogue infrastructure and many precision or particular settings are still made using test cards or similar patterns in conjunction with oscilloscopes and vectorscopes. Similar devices were also invented for digital video equipments.

Bizarrely, in Italy during the boom of the private TV stations (1980-1990) there, the appearance of a test pattern with an ID sign (from a monoscope, coloured bars with a superimposed logo or even a low-cost home computer) led in point of fact to the occupation of frequency. In the absence of a definitive assignment of TV channels, the first stations which transmitted such patterns owned the allocation of the airwaves! A sort of 'airwaves jungle' was born in those times in Italy, often culminating in a 'real war' between two or more misused and unauthorized broadcasters!

4.6 Software based Test Card Generators

The electronic design of a monoscope generator using ordinary ICs was and still is not an elementary task. In undertaking the necessary calculations, if just for generating some simple patterns and colour bars we have employed altogether more than twenty integrated circuits (included SPG625's devices), a lot of passive & discrete components and made some calibrations by means of specific tools, you can imagine how many components were necessary in order to create an elaborate figure such as the monoscope and how much accuracy is needed in its construction and calibration!

However properly skilled programmer could nowadays command a PIC microcontroller to generate an appropriate RGB+Sync signal containing an equivalent monoscope drawing ready to drive a PAL, NTSC or SECAM encoder with sync, blanking and pattern timings by following all the previous guidelines.

Another solution would be to upload into a non-volatile memory chip (EPROM, EEPROM, NVRAM, FLASH RAM, etc.) a high resolution image of the monoscope in such a way that every memory's cell contains a digitized point (pixel) of the monoscope itself. This image could be extracted by reading all the memorized information by scanning the memory using two binary counters driven by a master clock and synchronized by horizontal & vertical syncs (Fig. 28).

Many monoscope's images are available on the internet for uploading. The best respond to the 4:3 aspect ratio criteria and full height and width interlaced screen, i.e. a colour picture of 768x576 pixels.

Chapter 4. Pattern & Monoscope Generators

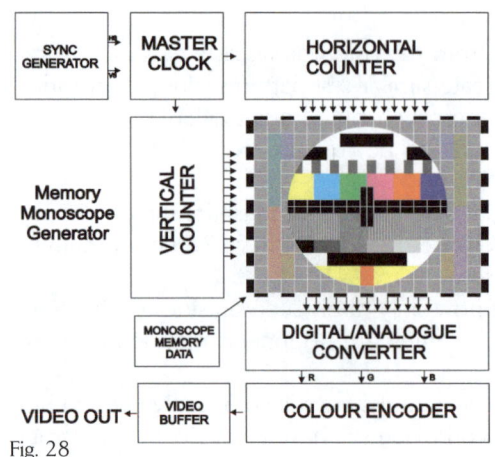

Fig. 28

Unfortunately this is not yet such a practicable way.

Since into a 52 µs active picture line have to be accommodated 768 points, every pixel will be about $52*10^{-6}/768 \approx 67.7$ ns wide which should be the total time needed for the memory addressing circuitry to select a single cell, clean, stabilize, D/A convert and present its data at output to finally drive the colour encoder together with the remaining cells' informations to get the monoscope.

Actually no commercially available non-volatile memory is capable of such read performance! Of course we can employ a more expensive high speed static or dynamic memory, which could support such read access time, but this would imply a long battery lifetime or backup supply otherwise on every blackout we should reprogram the memory! On the internet some tricky solutions appeared for counteracting the low speeds of non-volatile memory among which are smart multiplex data addressing circuitry and upload of a much lower resolution monoscope image.

Later we will introduce an elegant solution which creates a PAL high resolution video colour logo (and then a monoscope too!) by programming a special low cost high speed non-volatile static memory called NVSRAM.

A third method is based upon a compromise solution.

Quite recently, on the internet appeared some computer programs which can create (as diagrams viewable on the computer screen) all the test patterns listed above and much more. These software tools are absolutely freely downloadable from the *World Wide Web* and allow one to check the grade standard of a computer monitor set which, for its particular design and use in professional confines, needs accurate default calibrations as well as for fixing drift or repairing as opposed to the more tolerant adjustments of home TV sets. In fact in a computer monitor the features of frequency response, RGB convergence beams and focus (among many others) are very important factors because, for example, a CAD designer might need to distinguish between two very narrow points or tracks of his work.

Furthermore a computer monitor locks to frequencies outside of standard TV horizontal sync (ranging from 31 KHz to 100 KHz and sometimes more) and a multi-frequency scan 'classic' monoscope built with ordinary components could be very much harder to design even as expensive to buy for a TV repairer. So some code programmers have been motivated to create software-based test card pattern generators, some of which are fully customizable.

Herman J. S. Aben from Philips Electronics in 1997 released the latest version (v.3.11) of his freeware program simply named *Testpattern generator* which runs under Windows™. It can be used in all video card's graphic resolution modes also in true colour mode (24-bit), windowed or full screen as well as screen aspect ratios of 4:3 and 16:9 (Fig. 29).

In the same year the Dutch *Peter den Bak* released on the web his freeware *Testbeeld* (Fig. 30) able to reproduce a quite precise PM5544 monoscope design in addition to circle, grid, greyscale and colour bars patterns. The captions inside the top and bottom black boxes can be changed or erased totally.

4.6 Software based Test Card Generators

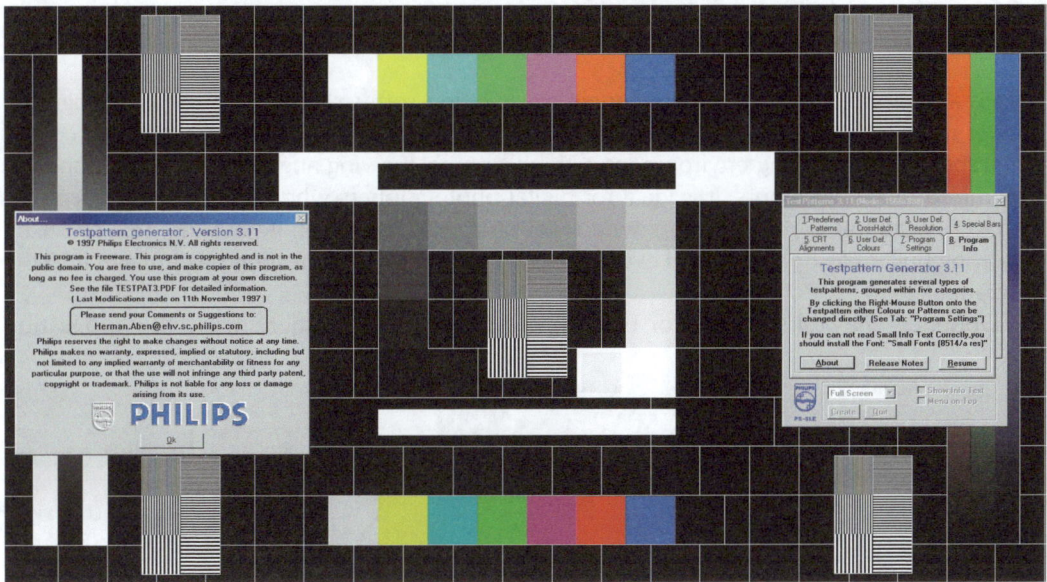

Fig. 29

Spectracal (www.spectracal.com) released its own free software test pattern called *CalM-AN™ HTPC Pattern Generator*, downloadable from its website but it needs a free registration. The software house provides also some ISO images to be burned with a Blue-Ray DVD recorder for playback on compatible Blue-Ray DVD players for DVD unit calibrations as well as standard or high definition TV sets for examination, since DVD support contains a lot of test patterns both in standard and high definition TV.

Calibration Aider from *Imaging Associates* (www.imagingassociates.com.au/color/testpatterns.

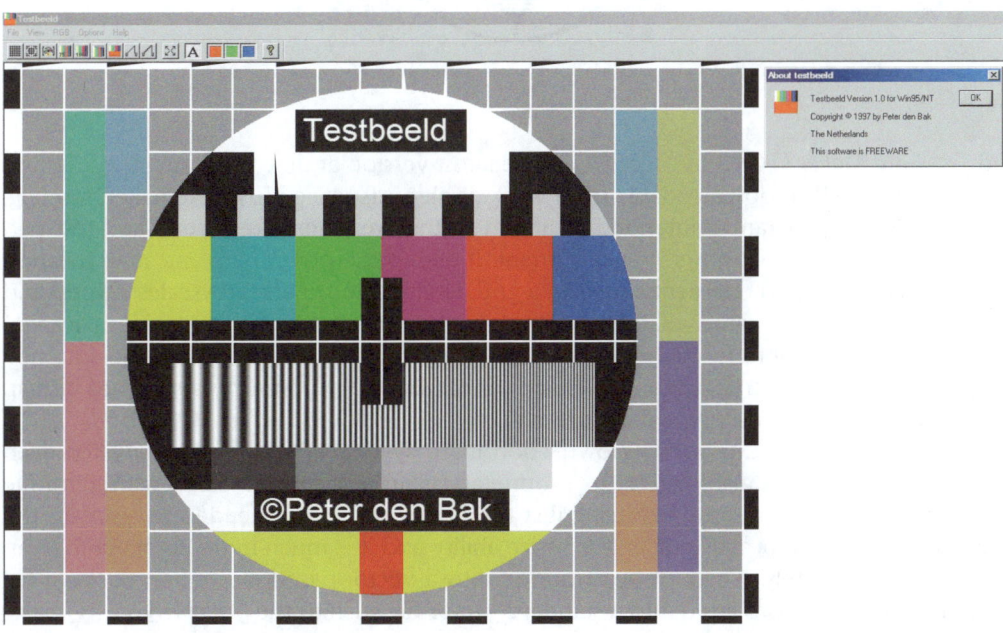

Fig. 30

Chapter 4. Pattern & Monoscope Generators

jspx) is a platform-independent free monitor calibration software based on the Java language. It helps users to realize the full potential of their computer display device by creating a consistent, reproducible viewing environment with an optimal viewing angle, screen resolution, and maximized range of colours. It is ideal for digital photographers who need to set up a colour-managed workflow, or home users who want to display digital photos to the best effect by displaying built-in test patterns and even allowing the users to import their own images into the program.

Doubtless *the best* software for building test patterns is a rather smart little application

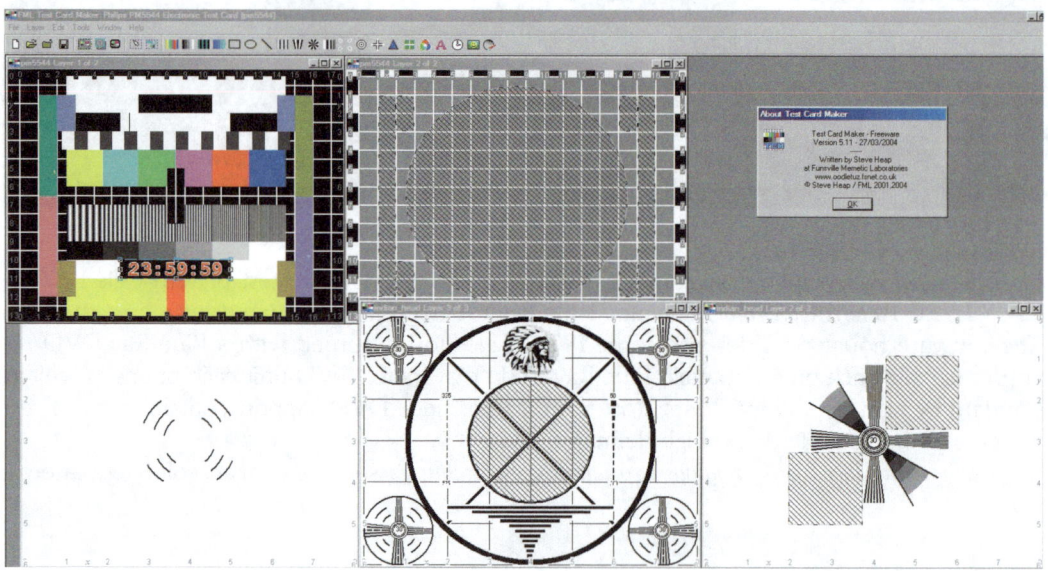

Fig. 31

called *Test Card Maker* (*TCM*) by *Steve Heap*. The latest version at time of writing is v.5.11 and dated 27/03/2004. It is downloadable from www.oodletuz.fsnet.co.uk/soft/tcmaker.htm (Fig. 31). This freeware program is intended as an aid to the creation of test cards and test patterns on the PC under Windows™ environment. It allows the positioning and manipulation of many basic test pattern elements, such as a grid, circle, colour bars, greyscales, sweep and multi-frequency gratings (rather than having to draw them manually with a graphics package) within a size-customizable multilayer work area. The drawings so created can be softened through a 3-level anti-aliasing tool and even anti-ringed before exporting as a standard bitmap uncompressed or JPEG-compressed picture file.

However because TCM uses its own coordinate system which you specify for each pattern, the program can also store in its proprietary plain text format (along with a *.tcd* file extension) the defined patterns by saving just the instruction codes needed to recreate the image. As a bonus this format retains true resize ability and it is much more compact than an image file. Unfortunately it cannot be imported into a vector-based illustration package for further elaboration or converted into a standard vector file such as the *Scalable Vector Graphic*

(*SVG*) or Adobe's <u>P</u>ortable <u>D</u>ocument <u>F</u>ormat (*PDF*) file.

To design a test pattern you have four line frequency options (625x25, 525x30, 405x25 and 819x25 lines x frames per second) as well as five aspect ratios (4:3, 14:9, 16:9, 5:4 and 1:1). The included specific design tools can be fully parameterized to respond to the needs of myriad monoscope and pattern varieties. The designing possibilities with this stand-alone program are impressive: it also includes the possibility of inserting in an area a real-time clock synchronized with the computer clock, a customizable text tool and it can even import external images. A huge collection of TCD files can be found on www.oodletuz.fsnet.co.uk/tcd, some of those are perfect reproductions of notorious TV and commercial monoscopes and test cards.

Test Pattern Maker 1.0 is probably <u>the</u> unique free utility for generating video and audio test patterns which runs under *MacOS 9*^{TM} as well as under *OS X*^{TM} in Classic mode. It is downloadable from *Synthetic Aperture*'s website (www.synthetic-ap.com) and it generates several video test patterns among which colour bars, 5 and 10 luma steps, luma ramp, grid convergence and others as well as accurate sine-wave audio test tones.

4.7 Two final remarks

In this chapter we have spoken about the generation of 'artificial' images by building hard-wired, simple, time-variable, repetitive and complex patterns. The software-based patterns and monoscopes are useful only for calibration of a computer monitor's due to the elevated scan frequencies of actual computer's VGA graphic cards and their particular 15-pins VGA or 29-pins DVI socket, thus they are incompatible with NTSC or CCIR625 devices.

However it is possible with some expediency to take advantage of the images generated via a personal computer to employ them on standard TV equipment. The most immediate way is to substitute the VGA video card with one with an embedded video or S-Video output port (PAL or NTSC), in addition to the standard VGA/DVI sockets. Through the video card's preferences setup it is possible to enable such a video port and watch the result on a standard TV set via SCART, RCA or S-Video mini-Din inputs. There is the collateral effect of needing a whole personal computer to be operational during all the TV needs, but with the advantage of obviating the need to acquire an expensive and self-limited TV digital test pattern and monoscope generator.

Unfortunately the video quality of inexpensive VGA cards is unambitious. Even the most professional VGA boards suffer from poor video output quality due overall to the scan conversion from a VGA to the TV standard frequency. It is possible to purchase of course an external broadcast-quality (and very expensive) VGA to PAL/NTSC Scan Converter, but you still need a PC to display the patterns.

However better results are possible if you use a computer with a native NTSC/PAL compatible composite video output (undergoing video encoding, either Y/C or RGB too) such as the early home computers (branded Sinclair, Commodore, BBC Micro, Acorn, DAI, IBM PCs and its clones equipped with EGA or CGA video cards, and others). The Commodore's Amiga computer could even 'switch on the fly' via software between NTSC and CCIR625 systems! I myself have drawn patterns and monoscopes selectable by keyboard for some private TV stations!

A smart, cheap and quasi-broadcast solution is based upon a DVD disk inside which we

Chapter 4. Pattern & Monoscope Generators

have previously authored and burned all the monoscopes, patterns, figures and pictures we have managed to design with Test Card Maker (or other tools) and saved as bitmap files. Explaining how to create this test DVD disk is beyond this book's scope; however this approach can quickly solve the issue by surfing on the DVD menu to choose any test pattern we want to visualize onto a display device screen, record on tape or broadcast. By connecting the video, S-Video, YUV/RGB components, and even VGA or HDMI (if available) outputs of a household (quite cheap) DVD Player to virtually any television (or even via VGA out to a computer monitor set) a video recorder or a television transmitter, we have realized the 'dream' to enjoy a multitude of multiscan (VGA & HDMI) and multisystem (PAL & NTSC) customized monoscopes and test cards without going bankrupt to buy equivalent commercial equipment!

Previously we have spoken about the relationship between the colour subcarrier frequency (F_{sc}) and the horizontal line frequency (F^H) of the system being considered.

An aspect of colour television systems that can appear difficult and obscure to explain is how to get such a relationship using electronic circuits. Unfortunately in all available diagrams regarding colour encoders the subcarrier oscillator is always left 'free' to run unlocked from line frequency. This is due to the complexity of achieving the full PAL subcarrier to line frequency relationship when designing a complete SPG which includes a composite video output and its relative circuitry.

As we know the PAL colour subcarrier frequency F_{sc} is given by the formula:

$F_{sc} = (1135/4) F^H + (1/625) F^H = 4.43361875$ MHz,

where F^H is 15,625 Hz,

or by the equivalent formula:

$F_{sc} = 283.75 F^H + 25 Hz = 4.43361875$ MHz.

The most difficult part in achieving the full PAL colour lock is the final 25Hz offset. However, in most cases the unlocked subcarrier does not cause any trouble to colour equipment which receive such a sub-standard colour signal. The reason is because these video facilities are designed to be much more tolerant to an uncorrected frequency relationship between subcarrier and line sync and lesser to the inaccuracy and instability of horizontal and subcarrier frequencies. The collateral side of this method is a visible cross interference which causes a crawling 'serrated' pattern primarily visible around coloured edges which slows down or accelerates randomly **but only when we work with a video composite signal**. When we work in S-Video (Y/C) or YUV/RGB components, the cross colour artefacts are invisible because they are absent.

4.7 Two final remarks

[1] Carol Hersee was represented in the 2006-2007 television series Life on Mars, portrayed by Rafaella Hutchinson in season one and Harriet Rogers in season two. Appearing to the protagonist in short visions, she would often taunt the main character Sam Tyler and sometimes he scared a lot. She was the last character seen in the series, when miming by turning a switch on the side of the TV screen the image disappeared like an old television.

5. Television Display Systems

Since the dawning of television, nobody could imagine the extraordinary *Evolution* of both display and capture devices, including also in the term *Evolution* a quite endless miniaturization of relative electronic circuits.

Taking you back to the '30s and '40s when relevant technology placed hundreds and hundreds of components in several rooms for televising something approaching 'acceptable' and nowadays, by putting billions of very advanced equivalent components in a space the size of a hand, we are able to create a quality of television reflecting the real world.

Besides the electronics subsystems of those times resulted in large build units equipped with fragile, unreliable and hot thermionic valves; their dimensions close to refrigerators and their heat dissipation close to wood-fired pizza ovens!

Furthermore, earlier pickup and (especially) picture devices were potentially dangerous both for their inventors and users due to the use of a technology based on vacuum glass tube. They could unexpectedly implode and seriously wound users, and to the extra high voltages needed for their functioning could cause electrical shocks while experimenting.

New technology introductions, to some degree, are forced by manufacturers who have to continually find something else better to sell, i.e. the next wave. The first picture and pickup devices produced unimpressive results, but goaded the researchers on to improve the technology. Here we are today, in the microprocessor age, with surface-mounted components soldered over flexible printed circuit boards, very flat display screens hung on walls and titchy solid-state cameras installed in mobile phones!

The advanced display systems improve on previous techniques primarily using the resources of human vision. The main goal of an advanced display is to improve the visual field occupied by the video image. In many applications the need has been for bigger, wider images that are intended to be viewed more closely towards conventional home cinema. In other applications, the need has been for a more miniaturised display to be used for specialized purposes where the viewer expects the image must have proportionally finer detail and sharper lines.

These are the display devices which followed this *Evolution* that their inventors or producers called in various modes in the context of those times: televisor, kinescope, CRT, FED, PDP, video wall, etc. These are only names of an evolutionary process started a hundred years ago.

5.1 A little bit of history...

It could seem absurd, but the first invention was a device that permitted one 'to see something of electric' on a screen and later another device which created a 'video' signal to be shown somehow on an 'electronic screen'.

The former picture device was a 'visual' application of *Thomas Alva Edison*'s electric lamp obtained through numerous successive improvements.

In 1884 Edison was investigating the reason why, after hours of operation the internals of his vacuum glass lamp became quite black thereby reducing the initial brightness. In the course of different tests to eliminate the inconvenience of that negative phenomenon, he tried to insert above the filament a thin sheet of metal, i.e. a sort of shield collector for the impurities released by the filament during incandescence.

5.1 A little bit of history...

This system did not eliminate the problem of glass ampoule blackening but he discovered a new phenomenon of great importance in the following years: the transfer of electrons (and therefore of electricity) through the air vacuum of the lamp. He discovered, in other words, that practically the filament bulb, without any contact, emitted a stream of electrons, or an electric current, towards the metallic foil applied in front of the filament itself.

Also he ascertained that by connecting a battery (in those times Volta's battery was the only electric source available) to a lamp modified with the metal plate foil, and having the negative pole on the filament and the positive pole on the foil, a passage of current occurred through the vacuum without a normal connection of copper or metal. Furthermore he ascertained that, by reversing the polarity of the connection to battery, the passage of current ceased completely. The same lack of electron transfer happened by turning off the filament lamp. Edison patented what he found, calling this effect the *Edison effect*. But he did not understand the underlying physics or the potential value of the discovery, although the thermionic emission effect was originally reported in 1873 by *Frederick Guthrie*.

However, the research with electron's transfer via evacuated tubes had started many years before that by scientists including *Johann Heinrich Wilhelm Geißler, Sir William Crookes, Eugen Goldstein, Nikola Tesla, Johann Wilhelm Hittorf* and many others, who had worked with high voltage accelerated 'cold' tubes, i.e. without an emitting heater.

Many years later the scientist *John Ambrose Fleming* further investigated on Edison's discovery and wanted to understand the principles of the Edison effect. Fleming found that when a filament immersed in a vacuum becomes incandescent, the electrons (of negative charge) which revolved around their atomic nucleuses escaped from their orbits in proportion to the filament's temperature. The electrons quickly were attracted by the metal plate with positive polarity which he called the *Plate* or *Anode* while the device he called a *Fleming Oscillation Valve* or *Vacuum Tube Diode* or *Kenotron*: this is considered to be the start of a technological revolution.

A couple of years later another scientist, *Lee De Forest*, by studying the discovery of Edison, ascertained that the electron stream (and therefore the electric current circulating between filament and plate in the vacuum tube) could be increased or decreased by inserting another electrode near the filament and equidistant from the plate. The negative electrons emitted by the incandescent filament were attracted by the positive polarity of the plate and also had to pass through this new electrode which could control the quantity of the flow and hence the electric current. This new electrode was called, for its precise function, the *Control Grid a*nd the tube was called the *Triode*.

Over years, the valve of Edison, Fleming and Forest, was added with further electrodes, each with its own function, such as *Screen Grid*, *Suppressor Grid* and finally an independent electron emitter, the *Cathode*, i.e. a metal shield that almost entirely covered the incandescent filament yet allowing the transfer of its heat, thus removing the low-frequency interference of alternate current (50 or 60Hz) present in the filament when powered from the electricity mains, which had meanwhile established. The relative tubes were known subsequently as the *Tetrodes* and *Pentodes*, and even with more electrodes also as the *Hexodes, Heptode* or *Octodes*.

Surely *Karl Ferdinand Braun* (Fig. 1) benefited most from his colleagues' studies upon the electron-emitting hot and cold cathode

Fig. 1

electrode tubes, or simply *Cathode Ray Tubes (CRT)*, by refining the electron emission into a beam and placing other electrodes on the tube's sides which permitted deflection of the emission in any 2D direction over a flat target covered by a fluorescent material: he had invented the *Braun Tube*, then called the *Oscilloscope*. For these studies he received a Nobel Prize in physics in 1909, in sharing this with Guglielmo Marconi. The principle of the Braun tube, i.e. moving an electron beam by means of an alternating voltage, is the principle on which all television picture tubes operated during all 20th Century. Lately, CRT production is quickly being phased out in favour of other display technologies which we will discuss later.

The deflection electrodes, called the *Deflection Yokes*, continued to be employed in tube-based pickup cameras until the '90s when finally electronics manufacturers developed fully solid state image capture chips, manufactureable in large scale which ultimately dominated.

5.2 Display devices

Braun's invention brought an authentic revolution in television especially when the CRT outclassed every mechanical television system due overall to its relative greater reliability and improvements over the years.

Nonetheless by the early '30s it was clear that Baird's mechanical television display could never produce the picture quality required for commercial success. The electronic television which required a CRT as picture tube for displaying televised images was relatively easy to develop, but the emerging electronic television was delayed for years until a suitable electronic camera tube could be developed.

Today television screens are an integral part of our lives.

Just consider for a moment, there are TV screens in the most unthinkable and diverse places, no longer relegated to mere projectors for moving images within the home's walls. We can find TV sets in surgery rooms, in surveillance installations for power plants, in railway stations and airports, in libraries and banks and even in the orbiting space station! All these TV screens display data, waveforms, informations, images, movies, animations, concerts, live events, etc. to satisfy our panchromatic eyes. It is hard to believe that not long ago televisions were a matter of a dream!

Looking back at the history of the development of a colour television system with backward compatibility with B&W TV sets, we can see that progress in many fields of technology, including television, depended directly on the progress of a number of adjacent sciences and technologies.

On the other hand, the development of colour television beneficially influenced other technologies, inciting scientists and engineers to new discoveries in upgrading existing devices or building new ones in other scientific disciplines too.

Nowadays CRTs have been increasingly replaced by digital colour displays that allow sophisticated views of audio and video signals, photos, vector signals, signal jitter and data analysis, although we can assume that the traditional CRTs have not disappeared entirely. The flexibility of digital technology, however, brings many new features, along with the replacement of CRT technology, which is also increasingly difficult to find spare parts for the oldest sets. As television sets and professional computer monitors, the industry is moving to more modern technologies and the benefits that the new colour displays afford.

The following two chapters are a cover the various invented and produced picture and pickup devices from early television days to our modern times... and beyond.

5.2 Display devices

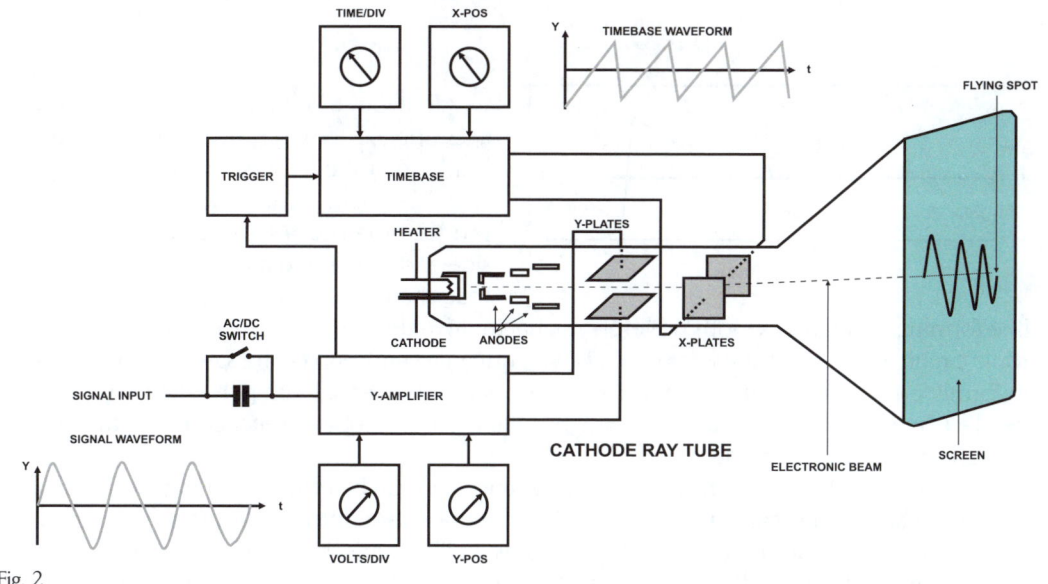

Fig. 2

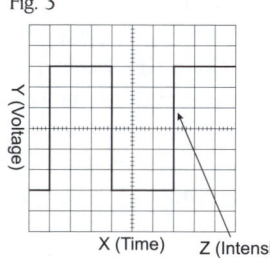

Fig. 3

Fig. 4

5.2.1 The Oscilloscope

Previously we have described the theory the finest details of the nature of the video signal. Now it is time to introduce a couple of instruments that will allow us to see these precise video waveforms and their formations. The most important and frequently taken-for-granted measuring instrument is the so called *Oscilloscope*.

An oscilloscope is a kind of 'programmable television set', although its size and shape are different. In fact its CRT-derived left-to-right scanning is very similar to that of a picture monitor, but we can manually change its frequency. However the vertical top-to-bottom scanning is not like a CRT but a representation of the voltage present at the input terminal. A typical functional diagram is shown in Fig. 2 whilst its common front panel is in Fig. 3.

The function of an oscilloscope is extremely simple: using an electronic beam striking on a sensitive screen it draws a *V/t graph*, i.e. a graph of *Voltage* against *time*, with voltage on the vertical or *Y-axis* and time on the horizontal or *X-axis*. In other words, the main purpose of an oscilloscope is to display the level of a signal relative to linear changes in time (Fig. 4) following the waveform variations of the input signal. Since the CRT produces a single dot of light, this dot must be swept constantly from left to right to create the illusion of a continuous solid line. In the resting position, without any signal to be measured or scanning, the beam would hit the screen centre, showing then a little fixed point. However, this is not convenient since there would simply appear a small glowing area at long time.

Chapter 5. Television Display Systems

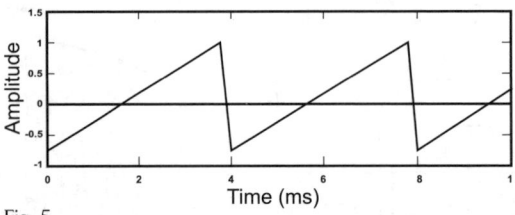

Fig. 5

To avoid this, there are two varying voltages applied to two special sets of electrodes inside the tube. One set comprises the *Horizontal Deflection Electrodes*, or *X-Plates*. These produce the beam's movement from side to side and are tied to a ramp voltage generator that changes continuously and linearly with time forming a sort of sawtooth waveform signal to drive the X-Plates. The ramp generator electronic circuit is called the *Time Base Oscillator*. During the ramp-up of the sawtooth, the beam is uniformly pushed from left to right across the front of the screen. During the sharp ramp-down, the electron beam returns rapidly from right to left, but the spot is switched off so that it does not appear on the screen (Fig. 5).

The 'slope' of the sawtooth varies with frequency and can be calibrated using the TIME/DIV control to change the scale of the X-axis. Dividing the screen into squares to form a graticule, the horizontal scale can be expressed in seconds, milliseconds, microseconds or nanoseconds per division (s/DIV, ms/DIV, µs/DIV, ns/DIV). Alternatively, if the squares are one centimetre wide per side, the horizontal range can be given as s/cm, ms/cm, µs/cm or ns/cm.

The signal which has to be displayed is connected to the input connector which will drive the *Y-Amplifier Stage*. An AC/DC switch selects a direct connection to the Y-Amplifier or a DC block of the signal so allowing only AC signals to go forward.

The Y-Amplifier is connected to a second set of electrodes, the *Vertical Deflection Electrodes*, or *Y-Plates*, so that it provides the Y-axis of the V/t graph. The gain of the Y-Amplifier can be adjusted using the VOLTS/DIV control, similarly to TIME/DIV, so that the resulting display is neither too small nor too large, but fits the screen and can be seen clearly. The vertical scale is usually given in V/DIV or mV/DIV or alternately in V/cm or mV/cm.

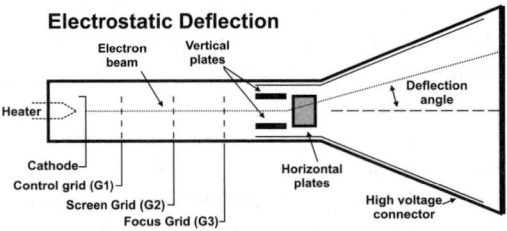

Fig. 6

Both deflection electrodes compose the so-called *XY Electrostatic Deflection Yokes*, utilized only for oscilloscopes or similar measuring instruments for its very accurate beam deflection as opposed to the less accurate *Magnetic Deflection Yokes* used in CRT picture tubes only (Fig. 6). The oscilloscope could also own a third axis, called *Z-Axis*, representing the intensity or brightness variations as the trace is formed on the display.

To provide a more stable track visualization, the oscilloscope has another important circuit called the *Trigger*. When using triggering, the timebase oscillator waits for a specified event before drawing the trace, usually achieved by a specified threshold voltage in the waveform input signal manually selected by the user or automatically by the first rising or falling waveform edge.

The resulting effect is to permanently synchronize the input signal, preventing horizontal drift of the track. In this way you can display periodic signals, such as sine and square waves, or complex and aperiodic signals such as composite video or single pulses. Without this feature, the displayed waveform would be constantly flickering and roaming, because every time the scan starts, the waveform track would probably be in an indefinite part of its cycle.

The oscilloscope is also a quite accurate multimeter and frequency counter but with an important difference from those real instruments. In fact it can measure the amplitude of a complex signal among its peaks and also the length of periods (and therefore frequencies) when the signal under observation is composed of two overlapped waveforms of different frequencies such as just composite video synchronisms.

Early and traditional oscilloscopes are completely analogue devices. Many simple and cheap analogue oscilloscopes have typical bandwidth of 20 MHz or less, but some better ones go to 100 MHz or higher.

Very early oscilloscopes employed huge CRTs due to the difficulties in those times to industrially produce small glass tubes. The analogue oscilloscope's viewing screen was and is generally a rectangular area on which is interleaved a transparent glass or plastic cover used as to offer protection to the tube from accidental bumps and on which is marked the graticule scale. However all the oscilloscope's CRTs have something like the shape of a cone whose bottom is usually a 2-inches diameter viewing screen for the cheapest instruments or 4 inches (or even more) for professional ones.

Most sophisticated analogue oscilloscopes have 2-inputs with dual traces which share the same timebase and triggering (yet with two separate Y-amplifiers) but switch the dot between one channel and the other either on alternate sweeps (ALT mode) or many times per sweep (CHOP mode). The quite expensive dual-timebase oscilloscope has two triggering systems so that the two signals can be viewed on different time axes.

Other designs include multichannel oscilloscopes that do not have multiple electron beams but rather they display only one dot using time-multiplexed channel switching.

In contrast to an analogue oscilloscope, more modern digital oscilloscopes use an analogue-to-digital converter (ADC) to convert the measured voltage into digital information. It acquires the waveform as a series of samples, and stores these samples until it accumulates enough samples to describe a waveform. The digital oscilloscope then re-assembles the waveform for display on the screen. The same waveform can also be stored into a memory and shown on its own or at a computer screen. The benefit of the digital technology is that the waveforms can be frozen in memory and then analyzed immediately or later, or even saved in an external memory. The digital oscilloscope viewer is a flat rectangular LCD screen of many different sizes and colours.

Although most people think that the oscilloscope is a self-contained instrument in a box, a new kind of tool is emerging consisting of an external analogue-to-digital converter connected via USB to a personal computer. This offers viewing, control interface, disc storage, network connection and sometimes the power supply. The most evident advantage is the relatively low cost in relation to its performance, assuming that the user owns already a PC. This makes the tools suitable for the education market, where PCs are common, but the equipment is often low budget.

The input connection impedance is around 1 MΩ in typical normal oscilloscopes and 50Ω in many high speed oscilloscopes. The signals to be measured pass through via one or more dedicated probes.

Chapter 5. Television Display Systems

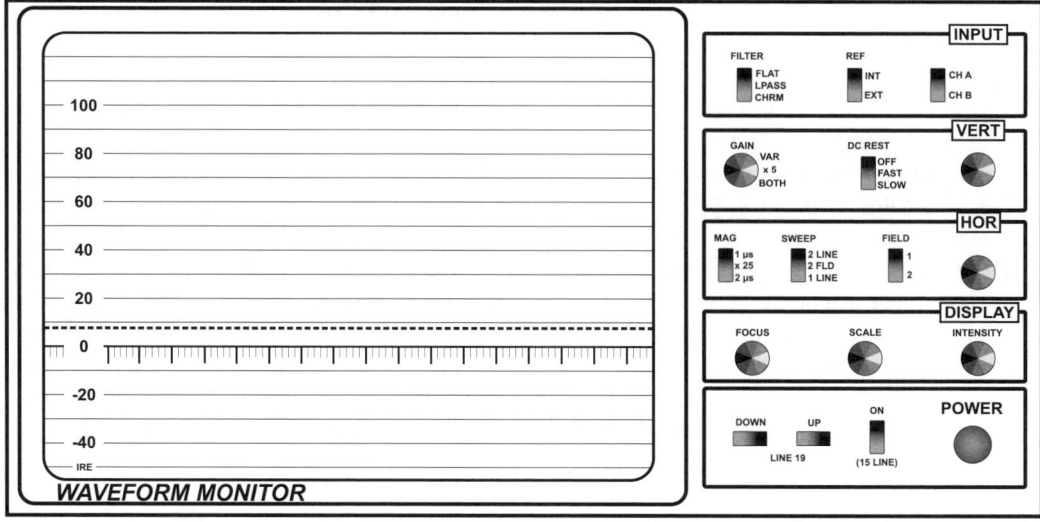

Fig. 7

5.2.2 The Waveform Monitor

A special type of oscilloscope especially used in television applications is the *Waveform Monitor*. It is calibrated to measure and display the level, or voltage, of a video signal against a graduated scale covering the screen (Fig. 7). It looks at luminance and other parts of the composite video signal as well as usual other key parts (sync, blanking, video, etc.) that are all required to make up a TV picture. More precisely, it is able to analyze the information from an entire frame or from just one or two lines of video. To do this a sync stripper circuit is used to isolate the sync pulses which feed a sweep circuit driving the X-axis. The incoming analogue video signal is filtered and amplified, and the resulting voltage is used to drive the Y-axis of a cathode ray tube, as a standard oscilloscope.

The Waveform Monitor's screen can accurately display and measure all of these video elements because it is divided into several horizontal lines representing 140 *IRE (Institute of Radio Engineers)* units. One IRE corresponds to 10 mV of video signal and therefore the ideal composite video signal ranges from 0 to 100 IRE units. For NTSC the actual minimum black level is 7.5 IRE units, while for PAL the black level corresponds to blanking level (i.e. 0 IRE on waveform monitor scale), and maximum white peak level is 100 IRE units. Obviously the sync pulses must drop below 0 IRE, typically -28 IRE for NTSC (or -30 for PAL) or even down to -40 IRE units.

Monitor waveforms have been a part of television since the first television black and white broadcast in the late 1930. For decades they helped technicians and engineers to adjust and understand the degradation of signals. The first units were effectively special-purpose oscilloscopes to measure the TV signal. When colour completely replaced B&W transmission in 1960, waveform monitors based on solid-state components have been developed, though still requiring a CRT as the display.

Today, with the advent of digital television and digital signal processing, waveform monitors

5.2 Display devices

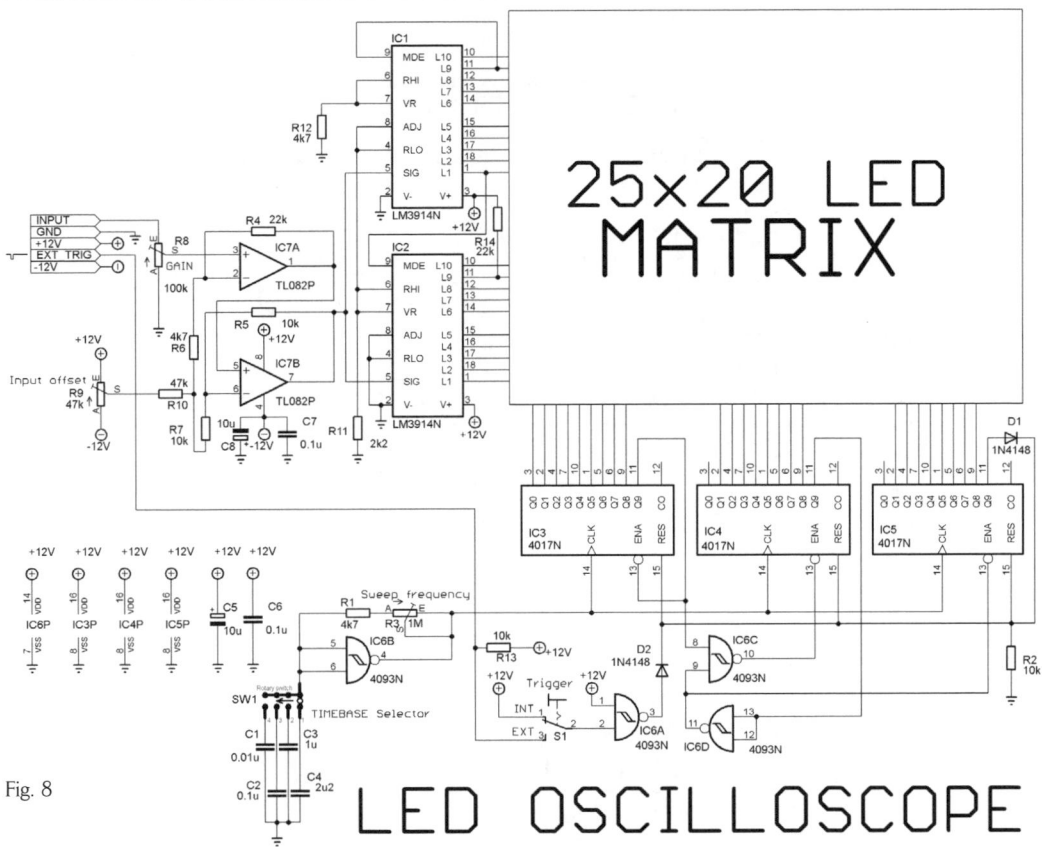

Fig. 8 LED OSCILLOSCOPE

have acquired many new features and capabilities. Modern waveform monitors include several additional operations, including display of the complete image just like a TV set; some ways to optimize the control range of colours; support for audio of a TV program; measuring the physical parameters of the serial-digital television formats; examining the serial digital layer protocol and for displaying data and metadata information as timecode and closed captions.

Modern waveform monitors and oscilloscopes have largely abandoned the old CRT technology. All new waveform monitors are based on a piece of circuit that duplicates the behaviour of a CRT display. This graphics hardware is called a *rasterizer* because it generates a raster signal which can be shown on a flat-panel liquid crystal display. The unit may even be sold without a display relying on the user to connect any normal TV set or even a VGA monitor to see and calibrate the video waveforms.

Some waveform monitor cards have also been developed for personal computers whose results are visualized onto computer screen altogether with the waveform monitor regulations.

5.2.3 A LED oscilloscope

In Fig. 8 there is shown the diagram of a funny oscilloscope offering a 25 rows x 20 columns LED matrix. This LED oscilloscope should be considered as 'experimental' and then evaluated 'as is' since the maximum frequency acceptable is 500-600 Hz, which is sufficient to see an intelligible waveform on the LED viewer. However it could be considered as a 'starting point' to be improved or extended guided by the following circuit's description.

Chapter 5. Television Display Systems

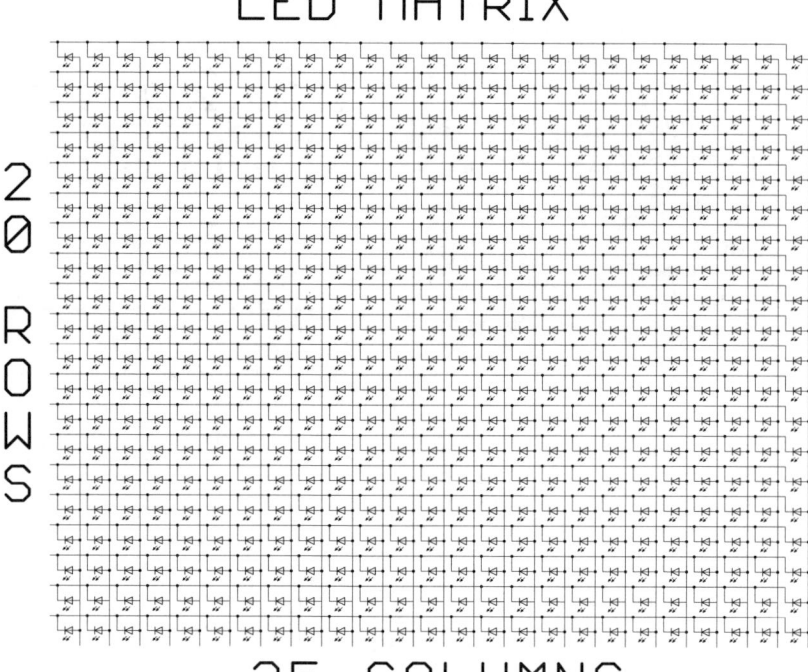

Fig. 9

Through seven ordinary ICs and a few passive components we can visualize on the LED matrix sinusoidal, trapezoidal or square waveforms and other simple figures. IC7 (TL082) is the input amplifier which drives the 'Y deflector' formed by IC1 & IC2 (2xLM3914), connected in cascade mode. The National's LM 3914 is an interesting chip designed to substitute the traditional VU-Meters by sensing analogue voltage levels. It furnishes ten outputs (active at '0' logical level) for driving 10 LEDs, providing thus a linear analogue display. The LM3914 can operate in two modes, dot or bar displaying mode; in our case both devices employed in this circuit are configured to turn one LED on only at any one time ('DOT' mode) because IC1 & IC2 rely on a half portion of the input signal. R8 is the Y gain calibration whilst R9 adjusts the input voltage offset.

IC3, IC4, IC5 (3x4017), IC6C & IC6D (two 2-inputs Schmitt NAND included in 4093) constitute a 25-output shift register fed by the oscillator IC6B (another 2-input Schmitt trigger NAND included in the 4093) used as TIMEBASE. Since the 4017 is a decade counter with 10 separate outputs (Q0 to Q9), one output of each goes high in turn as counting advances. The three counters own a common clock and reset line, so that for the first counter IC3 we have to use only the first nine outputs (Q0 to Q8), while for the second and the third counter only eight outputs of each counter (Q1 to Q8). In this way, also by means of IC6C & IC6D which enable the counters one by one in a precise sequence, we have achieved a 25-steps horizontal scanning system which represents the 'X deflector'. The last output of IC5 (Q9) via D1 will reset all the counters and the cycle reverts from the beginning.

Each active output of IC3, IC4 & IC5 is connected to all the anodes of a 20 LEDs column whose cathodes are connected each one to the outputs (representing the rows) of IC1 & IC2. All the remaining LEDs follow this configuration creating in such a way a typical matrix table (see Fig. 9). Since only a LED column is activated (i.e. going high logical level) in sequence and a single LED row will be enabled (going low logical level) according to the level input, the intersection point will produce the lighting of the correspondent LED. Our human persistence of vision does the rest in visualizing a continuous waveform on the LED matrix.

The timebase is a simple square wave oscillator made up of IC6B, a 2-input Schmitt trigger NAND included in IC6, whose range frequency can be changed via the rotary switch SW1 and swept with R3.

The last Schmitt trigger NAND IC6A enables the external triggering by means of the S1 position and negative-going pulses at EXT-TRIG input all at CMOS logic levels.

Long wires from the 4017 outputs to the LED anode common could cause oscillations. Depending on the severity of the problem, small value (1-10nF) decoupling capacitors from the LED anode common to the pin 2 of both IC1 & IC2 (GND) will damp the circuit. If the LED anode line wiring is inaccessible, often similar decoupling capacitors from pin 1 to pin 2 of IC1 & IC2 will be sufficient. If several LEDs light then oscillation or excessive noise is usually the problem. In cases where proper wiring and bypassing fail to stop oscillations, the power supply at pin 3 (always of IC1 & IC2) is excessively noisy and therefore it must be cleaned as much as possible. For a compact construction, SMS LEDs are preferable provided that they are of the same type and colour for the best performance. R11 & R12 program the current (I^{LED}) of each LED (and therefore its brightness) calculated with the formula ($I^{LED}=12.5/R^{LED}$), where R^{LED} is the resistor current for the lower 10 LEDs. The value resistor for the upper 10 LEDs can be calculated by multiplying the R^{LED} value by two. Actually they are computed to give around 6 mA per LED.

This circuit needs a dual voltage power supply of ±12V respect to ground. With reference to the LM3914's datasheets it is possible to test other features using BAR mode also.

5.2.4 The Vectorscope

Another specialized oscilloscope called the *Vectorscope* is often used in conjunction with a waveform monitor. Its task is to analyze and accurately measure the chrominance portion of a video signal.

Originally, these were separate devices but today it is quite common for the waveform monitor and vectorscope to be combined into a single unit that can switch between the two functions. Some units even allow for the two functions to be superimposed or they can be sold without a display, just like the new waveform monitors, being based on a rasterizer. In this case the user can connect a standard TV set or computer monitor to see the results.

This test instrument is used in combination with a colour bar generator for investigating amplitude and phase errors of the chrominance signals which represent the colour saturation and hue of the scene, respectively. This information is displayed on the vectorscope's screen in the form of a quite similar Lissajous pattern or 'dot runway landings', displayed by a vectorscope in a circular display (Fig. 10).

Hue tint is represented by the degrees of rotation from the reference point and saturation is represented by the distance from the centre of the circle. The circle's circumference is sub-

Chapter 5. Television Display Systems

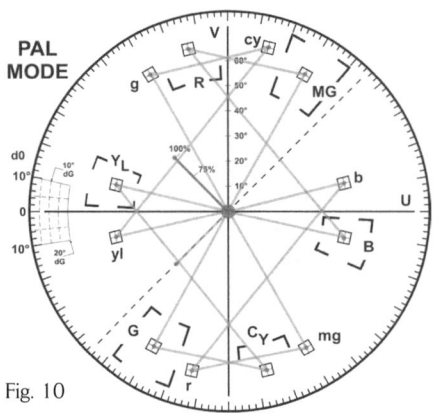

Fig. 10

divided in 360 parts, like a sort of goniometer. Each part represents one degree of the phase relationship between a colour and the subcarrier reference.

The centre of the circle is neutral, so that the closer the colour is to the centre, the less saturated it is tending towards almost white; the more distant it is from the centre, the more saturated or less neutral it is. A colour can be dark and very saturated or light and unsaturated. A black and white video signal will be represented by a point at the centre of the circle as well as any grey tone raster.

Inside the circle are also shown a series of small blocks, which indicate the position of the red, green, blue, yellow, cyan, and magenta colours of the colour bars signal used as reference.

Square brackets mark the position of each pure hue on the face of the vectorscope telling us precise information about the colour balance of video equipments.

A vectorscope is a quite expensive measuring instrument; for this reason it is extensively used only in TV stations or laboratories specialized in repairing or calibrating professional video cameras or similar equipments. A regular television set repairer generally would not need it.

5.2.5 The Baird's 'Radiovision', the first industrial mechanical television set

Previously we broadly stated that *John Logie Baird* was one of the real pioneers of television. His mechanical *Radiovision*, designed and drawn by himself, was manufactured by the *Plessey Company* in England and it was the most popular version to be available to retail from 1928 to 1934 in the United Kingdom, United States and Russia. It was especially purchased by British television amateurs in order to watch the regular broadcasts of the BBC & Baird Studios (via the *Baird Television*, the first 'independent' TV station of the history) then available.

However, the first sets sold commercially in 1928 in the United Kingdom and the United States, made by Baird himself, were simply radios having the addition of a vision device consisting of a neon lamp behind a electromechanical spinning disk (the *Nipkow Disk*, Fig. 11) holding a spiral of apertures. It produced an image the size of a postage stamp (6 cm in height and 2 cm wide), with 30 lines of definition magnified twice via an optical system and reproducing a few images per second. The images produced instead of white and black, were actually black and red because of the colour of neon gas in the lamp (Fig. 12).

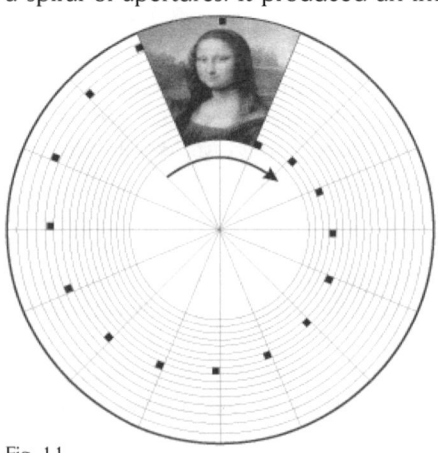

Fig. 11

It was a curious device that Baird built in a basement of the infamous Soho district, thanks to money received by the first sponsor of TV history, the film impresario *William Day*, captured with a classified on the *Times* magazine.

The low definition resulted in a video bandwidth of less than 10 kHz, allowing pictures to be broadcast using an ordinary AM/MW or LW radio transmitter.

5.2 Display devices

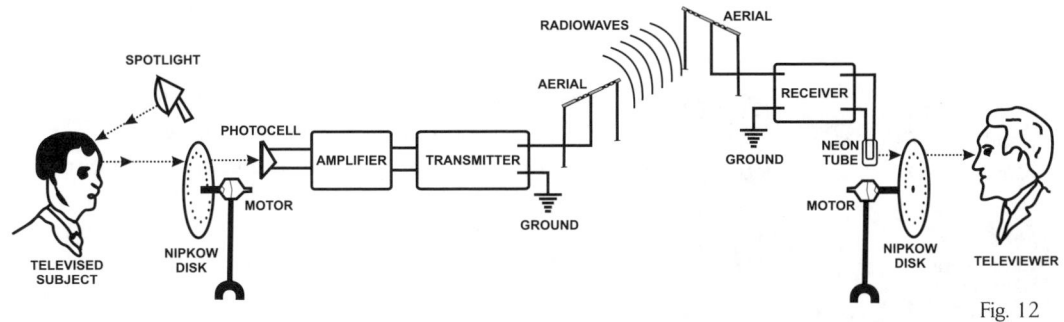

Fig. 12

The resolution soon improved to 60, 90, and 120 lines and then stabilized for a while on 180 lines (Germany, France) or 240 lines (England, the United States) around 1935. Scanning was always progressive, which means that all lines of the pictures were scanned sequentially in one frame.

About a thousand of the original Baird sets were built and sold at a price of just over eighteen pounds each for the British and Western Europeans, which represented an astronomical sum for those times. A variant was built also as a receiver set without the metal cabinet available for 'only' seven pounds and a version without the radio, known as a *Televisor*. Similar 'low-cost' DIY kits were much more popular, based on same design. Baird's rudimental units are considered the first mass-produced television sets and they are memorabilia now highly sought by collectors and museums because most people do not recognize it any longer simply as just a television set.

Since the scansion system was fully mechanical, the audible noise was terrifying and the pictures produced were barely discernible. However Baird had successfully demonstrated the capability of the mechanical television based upon Nipkow's invention since the first television broadcast in October 1925 and subsequent refinements.

5.2.6 The Cathode Ray Tube (CRT) or Kinescope

The <u>C</u>athode <u>R</u>ay <u>T</u>ube (CRT), more popularly called the *Braun tube* in Japan and some other countries after its inventor, the German scientist Professor *Karl Ferdinand Braun,* was and probably still is the most widely used display device which contributed to the television revolution.

It is generally accepted that Karl Ferdinand Braun developed the first controllable CRT in 1897, when he added alternating voltage to the device to allow it to send streams of electrons in a controlled manner from one side of the tube towards the other. In an industry in which development is so rapid, it is somewhat surprising that the technology behind monitors and televisions is over a hundred years old!

However Vladimir Kozmich Zworykin experimented and patented in November 1929 an improved Cathode Ray Tube that he named the *Kinescope*. It was not until the late 1940s that kinescopes were commercialized in the first television sets.

Although modern CRT monitors have been subject to improvements over the quality of the images, they still follow the same basic principles. The CRT has applications which range from portable colour television sets and giant screens to computer monitors.

The aim of CRT is to reproduce the original moving pictures (created by a video camera or similar device) as faithfully as possible. The fidelity of the representation depends on the type of unit, its properties and mode of operation. The oldest display unit is the traditional CRT which

Chapter 5. Television Display Systems

Fig. 13 - Alfred Norton Goldsmith (left) & Guglielmo Marconi (right)

remains a term of comparison by which all the other types of display devices have to be judged.

A great R&D effort went into the technology of colour CRT over many years to get to a screen diagonal of 35 inches (89 centimetres). Improvements have been made in the material of the screen, the technology of the electron gun and the electricity supply power saving. The colour CRT owns all the simple functioning methodologies of its B&W counterpart but for refinement and efficiency it is able to produce a colour image on a unique display. These colour tubes work with the principle of the *Shadow Mask* (a diaphragm with thousands of holes placed very close to the screen) originally patented by *Alfred Schroeder* in 1947 and brought to completion by Dr. *Alfred Norton Goldsmith* (Fig. 13) and his research team in the American *RCA Laboratories Ltd* in 1950. We re-affirm that the CRT technology is the oldest used in the television field! Nowadays the researches in display devices are moving in favour of alternate technologies which we will discuss later.

The early CRTs were rather clumsy, cumbersome, heavy and very much more expensive than current ones, but all own the essential features of colour tubes proposed for mass production with all the benefits that resulted. Indeed the earliest CRTs were so huge they were positioned vertically inside metallic or wooden cabinets with the target screen oriented on top. A mirror, which when the TV set was switched off covered the screen itself, reflected the generated images. The practical realization of a relatively compact and very inexpensive domestic TV receiver was (and still is) strictly related to the technological researches and developments over the years.

Today an increasing number of display devices are available including the latest LCD and plasma widescreens which are rapidly replacing older CRTs. However the CRT is a standard display device by which all other devices are compared. It is a diffused opinion among the professionals that CRT remains the best device in terms of picture quality.

In the CRT, electrons at high speed are organized into a single ray (or beam) and emitted by a gun (Fig. 14), usually a heated cathode. They are accelerated and focused by a sort of electronic lens, and proceed directly to a screen that serves as an anode with a positive charge. The screen, which is internally coated with a compound called the *phosphor* (which should not

Fig. 14

be confused with the chemical element *phosphorus*) gives a visible glow when hit by electrons. It emits a light whose colour is determined by the type of phosphors used, only one type for monochrome display and three types (representing the three primary colours) for a colour display.

The phosphors are known as type *P4* for black and white CRTs, and type *P22* for those in colour, according to the American nomenclature. The persistence of the radiation emitted after phosphors excitement is defined as *long, medium* or *short* depending on its duration in the order, respectively, of seconds, milliseconds or microseconds. A practical employment of a tube with high persistence phosphors was the *Commodore A2080* computer monitor, used thus to reduce flicker in interlaced modes. The monitor had a 'long afterglow' which left a slight temporary trail with moving objects generated by proprietary Amiga video computers displaying on screen in such a way a hi-res full-size interlaced images almost without flickering. Another example is the radar circular aviation tube with long persistence phosphor used in military aircrafts.

An *Extra High Tension* (*EHT*) anode voltage, typically 15-30 kV, is needed by the CRT to attract and accelerate the electrons. This EHT is generated by a voltage multiplier device known as the *Tripler* fed by a secondary winding of a special horizontal line transformer called a *Flyback* or simply *Line Output Transformer* (Fig. 15). The other high voltages necessary for the proper functioning of the CRT are obtained by some supplementary windings too as well as other low voltage DC supplies. An increase in the potential between the cathode and the anode causes more electrons to be emitted and vice versa.

The access to the various electrodes is obtained through metal pins placed in the back of the neck of the tube with the exception of the anode which is accessed through a special socket placed on the bulbed section of the tube usually very close to the front screen. The inside and the outside of the tube, connected to EHT and chassis respectively, are coated with a layer of graphite, and through the glass which separates them form a kind of EHT capacitor. This is obviously of a low value but due to the elevated charge it is important that this capacitance is completely discharged before handling the tube otherwise a very violent electric shock can be experienced, not lethal yet very dangerous!

In order to produce an intelligible image the electron beam is deflected in the horizontal and vertical directions. To achieve this two deflection plates move the electron beam, whose direction depends on both the polarity and the amount of electric potential present on the two plates. The first pair of plates deflects the beam in the vertical direction, while the second acts horizontally. When a receiver is tuned to a station, the scanning beam will be then synchronized to the incoming video signal generated via a remote camera or other similar device or, in other words, an internal sync generator is phase-locked to the incoming one generated from a *Sync Pulse Generator* (SPG), a device we have already widely discussed previously, located in the TV station and employed as a master reference signal. In modern CRTs the deflection is obtained by means of two pairs of magnetic coils, the *Deflection Yokes*, placed orthogonally and along the neck of the tube (Fig. 16). The relative deflection drive signals follow from the result of strong amplification from a sync locker.

The angle of deflection is an important feature of the CRT which refers to the angle through which the beam is deflected before hitting the screen phosphors. The most common deflection angle is 90 degrees, albeit 100° and even 113° (ex-

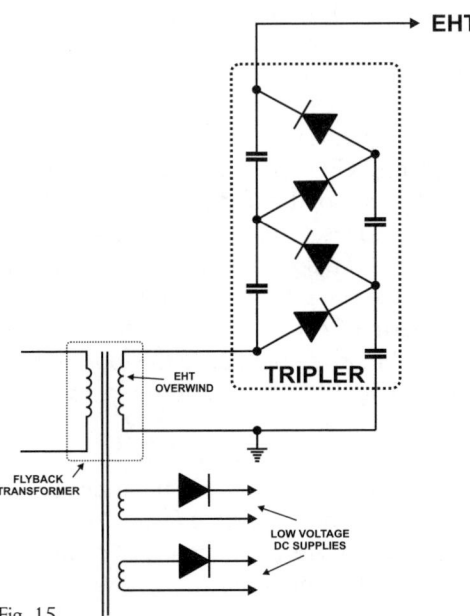

Fig. 15

Chapter 5. Television Display Systems

Fig. 16

treme limits) are widely used for large screens. The angle of deviation depends on the strength of the magnetic field created by the scanning coils, the speed of the beam of electrons accelerated by the EHT and the diameter of the neck of the tube.

The input signal is amplified and applied to the CRT's control grid whose polarity, and therefore the amount of electrons that go beyond to hit the screen, depends on the intensity of the signal received. This is a process known as the *Grid Beam Modulation*. Another type of modulation is the *Cathode Beam Modulation* where the voltage at the grid is fixed and that of the cathode varies with the video signal. This last technique outperformed because of its greater sensitivity with negative-going blanking pulses applied to the grid to switch the beam off during line and field retraces (Fig. 17).

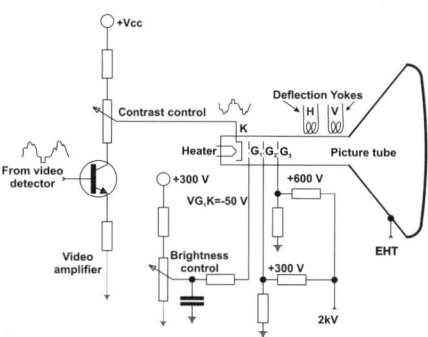

Fig. 17

Through the combined action of voltage scanning and that of the input signal, the electron beam draws on the CRT's screen a sequence of points that accurately reproduces the original image. Under weak radio reception the sync revealers try, as much as possible, to lock the incoming syncs, otherwise 'ripped', inconsistent, unstable or 'snowy' images will be generated. With no video signal the internal sync revealer generates horizontal and vertical pseudo-syncs to keep the raster screen someway 'opened' else a luminous dot at the screen centre will continuously appear with the risk of burning the phosphors permanently. In very old tube-based TV sets such bright dot popped up on CRTs after switching off because of the residual EHT CRT charge; alternatively a fault in the vertical or horizontal deflection stage caused the appearance of a bright horizontal or vertical line, respectively.

The length of the CRT diagonal determines its size, ranging from 1 to 35 inches (2.5 to 89 cm), although to obtain larger images such as for projecting CRT moving images onto a cinema screen some magnifying optical systems are used. The reason behind this strange oblique measurement is very simple. As the electron beam must be focused on the inner surface of the screen, the functional form for this purpose is still a section of a spherical surface. This avoids blurring and edge distortion of images from the centre to the sides of the screen and therefore the diameter of the original CRT screen featured it as a term of excellence (Fig. 18).

A quasi rectangular external non-magnetic (read plastic) mask camouflaged the top and bottom of the circle screen to give the 4:3 picture proportions (Fig. 19). Afterwards true rectangular CRT screens were produced.

However, the tube size does not respond to the actual display or visible screen size because it is normally partially covered by the front panel framework of the cabinet and the tube cannot project a whole image which entirely touches its edges. This is an important factor to keep in consideration while framing with a camera and especially with images generated via a computer. In both cases the operator can rely on a rectangular area, called the *Safe Area*, whose shape is

5.2 Display devices

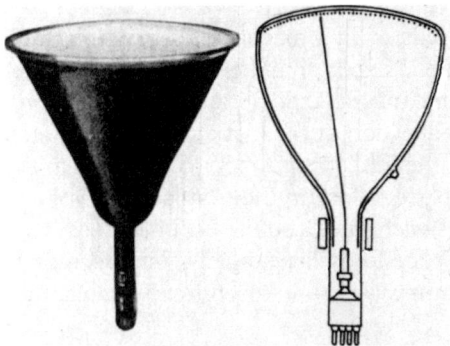

Fig. 18

Fig. 19 -Westinghouse H840CK15 TV set - Courtesy Early Television Foundation - *www.earlytelevision.org*

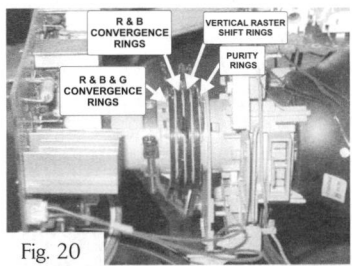

Fig. 20

Fig. 21

overlaid on the camera's viewfinder screen or on a preview monitor window of the computer program so to put the subject under observation at the centre of the scene. The area outside the Safe Area is called the *Overscan Area*, or briefly *Overscan* or *OS Area*.

A pure sawtooth scan waveform generator triggered by a horizontal line, similar to that found in an oscilloscope, drives the horizontal yoke. A similar waveform is produced for the vertical scan to feed the vertical yoke. Any non-linearity in the waveforms will produce a raster distortion called the *Pincushion*. Such distortion would be mostly evident on a flat CRT screen and may be reduced by a precision designing of the scan coils and by introducing an electronic non-linearity correction in the deflection stages for counterbalancing the defect, a process called the *S-Correction*.

Certain televisions exhibited symptoms such as poor alignment of geometry or bowing at the sides of the picture, similar to those caused by problems in the pincushion correction circuit. Improperly aligned geometry would clearly be apparent when viewing a crosshatch pattern such as those we have seen in the previous chapter. The user may complain about the effect on the picture while viewing letterbox movies or programs that display text information at the bottom of the screen, such as a stock market ticker. The convergence and pincushion adjustment procedures are more complex to perform manually for the flatter, more modern screen tubes, over a standard older spherical television CRT. Therefore in modern televisions there are more advanced on-screen serviceman-mode adjustments provided to facilitate these calibration procedures (Fig. 20 & 21).

The most parts of such adjustments are accessible simply by entering a specific key sequence on the associated IR remote controller and are setup through the remote keys themselves.

Colour CRTs generally have three separate cathode guns which for correct colour reproduction rely on the relative beams striking exactly their own phosphors only. The merit of success of this operation is termed the *Purity* and involves both the scan coils in terms of strength and direction for moving the beam.

Chapter 5. Television Display Systems

It is physically accomplished by the use of two-pole pairs of ring magnets and a circular disk magnet placed along the neck of the CRT. Any magnetic external influence, such as speaker magnet coils or mains transformers or even the terrestrial magnetic field, can permanently produce coloured splashes on the screen due to the magnetization of the mask causing unwanted deviations in the scanning beam. In order to avoid this, the CRT has a copper coil known as the *Degaussing Coil* wrapped on the bell of the tube and placed close to the front screen. TV and monitor sets automatically degauss their CRTs when switched on, before an image has been displayed. A degauss causes a magnetic field inside the tube to oscillate rapidly, with decreasing amplitude. This leaves the mask with a smallest and somewhat randomized residual field, thus removing the coloured splodges or the discolorations.

The three electron beams produce three separate rasters which must not only be rendered with the correct rectangular shape and no pincushion distortion, they must also be overlapped exactly. This relative calibration is called the *Convergence*, and is distinguished into two types, being *Static* and *Dynamic Convergence*. Static convergence is related to the central part of the screen whilst dynamic convergence covers the remainder, involving the continuously dynamic magnetic field produced by the scan coils to ensure convergence in the peripheral areas and display corners.

The *Delta-Gun Shadow Mask* kinescope was the first colour CRT produced in large scale employing the disposition at 120° of each one of the three guns as well as the dot phosphors in order to form an equilateral triangle or delta (Δ) from which its name derives.

Because of the presence of a shadow mask a large number of electrons fail to traverse the holes causing low efficiency and low brightness of the CRT which would need to be counteracted by increasing the EHT. For this reason (among many others including a complex and very skilled alignment work often performed for earliest CRTs even after transportation) delta gun CRTs are no longer used for domestic TV receivers. Yet for some time they continued to be produced for advanced computer monitors because of their high definition from the very fine-pitch shadow mask obtained through high performance construction and very precise tuning.

Later, the *In-Line Shadow Mask* technology was invented whose first application was developed by Sony who in 1969 introduced a tube known as the *Trinitron* (Fig. 22), a special tube derived from the *Chromatron* tube invented by Nobel prize-winner *Ernest Lawrence*.

Here the task of producing three electron beams is entrusted to a single electron gun with three cathodes arranged over a horizontal plane. The green cathode was placed centrally because the eye is most sensitive to this colour so that its placement on the tube axis does not commit any error of trajectory. The three beams are then focused using a single system of electrostatic lenses with a large aperture which because of its dimensions does not introduce any significant aberrations thus allowing a very significant focus of the three electronic beams across the entire surface of the screen.

The system also includes two electrodes for static convergence (equivalent to two optical prisms) that converge the three beams towards the centre of the screen, since these normally emerge divergent in the horizontal direction. Before arriving on the screen the electrons must pass through an *Aperture Grille*, which consists of a mask with a large number of continuous vertical fissures, created by chemical etching on a sheet made of a special alloy. The phosphors are deposited on the screen as terns of vertical stripes, aligned with the slits of the aperture grille. A series of non-reflective stripes are alternated with the phosphor stripes and are respon-

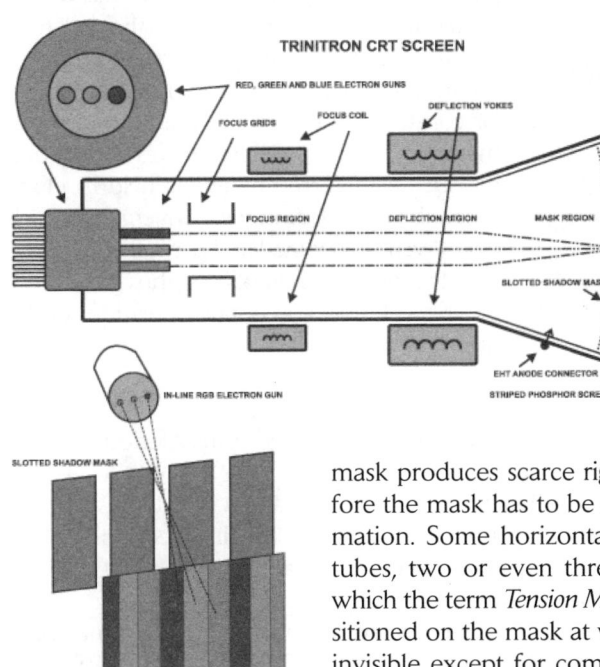

Fig. 22

sible for absorbing the rays of background or environment light. The convergence and purity in the vertical direction do not undergo detectable changes because of the arrangement of vertical stripes of phosphors; the same dynamic convergence is achieved in a much simpler way than with traditional shadow mask tubes.

However the Trinitron tube suffers from two main drawbacks. The first is that the manufacturing of the striped mask produces scarce rigidity in the vertical direction and therefore the mask has to be strained to prevent breakdown or deformation. Some horizontal, very tiny tension rods (one for small tubes, two or even three or more for large screens and from which the term *Tension Mask* is derived) were introduced to be positioned on the mask at various heights. These are normally quasi invisible except for computer monitors equipped with Trinitron tubes where the user might be bothered by such horizontal tiny rods that it could be misinterpreted as a tube flaw! The second drawback is the need for some dynamic convergence adjustments due to in-line disposition of the RGB guns, especially in wide-angle screens. Trinitron screens own the unusual characteristic of relating to a cylinder instead of a spherical sector.

Some improvements were achieved over the years and some spin-offs were developed from brands outside of Sony such as Mitsubishi and Eizo.

The Trinitron technology offered (and still offers) highly saturated colours and brightness, high contrast and a very pure white, despite its relatively old design.

Sony ceased Trinitron TV tube production in early 2008, albeit video monitors are the only remaining Trinitron-based sets to be produced at a low production rate and sold for the professional market.

Another type of in-line kinescope was introduced by Mullard/Philips and was known as the *Precision In-Line Shadow Mask* tube (Fig. 23), identified also with the letters *AX*. The three guns are placed adjacently and arranged aligned on the same horizontal plane, just like the Trinitron, as well as the screen phosphors in the form of striped triads. Such disposition brought many advantages among which the purity is no longer influenced by horizontal magnetic fields, such as that of the earth, and the need for vertical correction disappeared because the three beams always flowed on the same horizontal plane.

In addition both types of convergence are reduced to a relatively easy task of deflecting the two external beams to converge slightly toward the inside with the central beam and purity calibration was no longer needed by use of a self-converging mechanism. The mask is a robust

Chapter 5. Television Display Systems

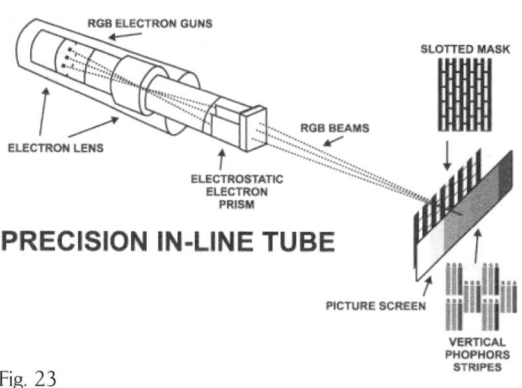

Fig. 23

grid with solid segmented vertical lines whose dimensions are in a ratio of 10 (height) to 1 (width). All these characteristics are obtained thanks to the combined action of the particular construction of kinescope and the conformation of special deflection yokes designed to produce a staggered magnetic field to eliminate the need for dynamic convergence. Such yokes were sold joined and permanently glued with the kinescope simplifying thus greatly aiding the assembly and testing operations of the tube.

5.2.7 Other types of CRTs

Prior to the mass-diffused colour CRTs based upon holed masks, some 'curious' colour cathode ray tubes were invented and developed.

Again *John Logie Baird*, pioneer in the mechanical televisions, was also a brilliant inventor in the colour electronic television field suggesting, around 1940, ideas for the first electronic colour television set by employing multiple cathode ray guns placed on separated axes.

Baird ascertained that if two 'teapot-like' tubes were mounted side by side using a common envelope having inside a quasi-transparent screen coated by suitable coloured materials, a two-colour tube could be produced. Two electronic guns would simultaneously strike the opposite sides of a 10-inch diameter clear mica screen, one side with a fluorescent dye sensitive to blue-green and the other side to orange-red. In this way the images would overlap to form a colour picture.

It was July 1943 when he called this device the *Telechrome* (Fig. 24).

Baird had also considered a tri-chromatic system which required a crinkled screen to somehow provide an additional surface for the third colour, but it is not clear whether the events in this direction have been made ever. He also designed a small Telechrome tube with a single electron beam perpendicular to the screen simplifying the scanning system. This has survived and is exhibited at the *National Museum of Photography, Film and Television* in Bradford, England.

The system worked with a definition of 600 lines with triple interlacing so it needed six scansions to complete a whole colour image. On August 16th, 1944, he provided a stunning demonstration for a group of journalists in his *Crescent Wood Road Laboratory*. This was the world's first demonstration of a wholly colour electronic television set using a single cathode ray tube. He later brought the resolution up to 1000 and even 1800 lines, three times the definition of a current colour TV system!

Fig. 24

5.2 Display devices

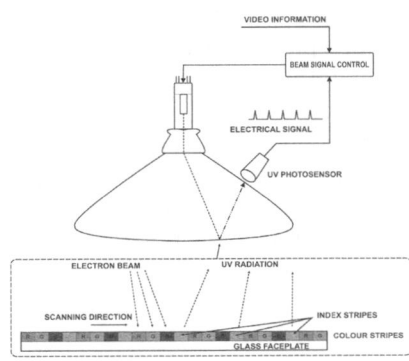

Fig. 25

John Baird died in 1946 at the early age of 58 years leaving his colour work unfinished and, as it has turned out, largely unknown and unfiled.

Another type of colour television cathode ray tube was the so called *Beam-Index Tube*, or *Apple Tube* as its designer and producer *Philco* (a <u>co</u>mpany from <u>Phi</u>ladelphia, in the United States) codenamed it (Fig. 25).

It used a series of coloured strips of phosphors and an active timing feedback, instead of a 'mechanical' focussing system developed by RCA for its shadow mask, so offering images much more bright. In addition to this a single electron gun rather than three, made its construction easier and it needed lower power. However, despite several years of development the sets were never able to be produced at sufficiently low cost and Philco finally froze the project for a while.

The development was not finished there, however. The system was reintroduced with

Fig. 26

the name *Unitron* (Fig. 26) in 1970. Substantial changes in electronics intervened during the last decade that significantly lowered the costs of implementing the system, until it became competitive with conventional CRT sets. Several Japanese companies have used the system for a variety of specialized purposes; the best known is the Sony *Indextron* series. The system was also used in the military for its ability to reject outside interference.

The tube showed images by illuminating vertical strips of coloured phosphors prepared in a red-green-blue sequence. A single scanning electron gun was used to properly hit the strips, and the strength of the beam was modulated to produce different colours.

Each RGB pattern was followed by a single strip of ultraviolet phosphor on the inner face of the tube, where the light is not visible to the viewer. The light emerging from this strip was captured by a photodiode placed on the outside of the tube whose signal was amplified and fed into the colour decoder to be electrically subtracted. This resulted in a phase difference that advanced or retarded the exploration beam by its occurrence on the horizontal deflection yoke. In this way, the index system would adjust the scanning timing on the fly in order to ensure a proper colour reproduction.

In order to receive a signal strong enough to index with, the beam had to be left on at all times, which reduced contrast ratio in relation to conventional tubes. Nowadays such types of tubes are quite obsolete.

5.2.8 Field Emission Display (FED)

Since the research on traditional CRT technology was practically finished, *Silicon Video Corporation*, in 1991, started to develop a new type of tiny CRT, called the *ThinCRT*, in order to create a flat panel display using large-area field emission electron sources to provide electrons striking on coloured phosphors so to produce a colour image. The field emission display had been invented in the 1970s followed by hundreds of patents on this technology (Fig. 27).

Chapter 5. Television Display Systems

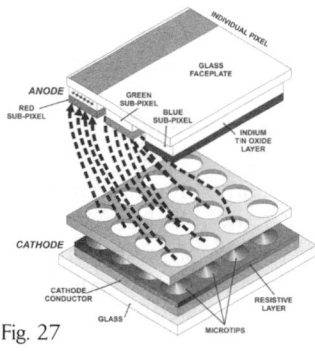

Fig. 27

The flat display consists essentially of a matrix of individual nanoscopic electron guns, each one producing a single sub-pixel, arranged in triads to form a typical RGB mosaic. The basic technology is similar to a CRT for which with FED were combined well-known features such as high contrast deep blacks and very fast response times (or almost absent motion blur) with the packaging advantages and the absence of magnetic deflection of other flat panel technologies such as Liquid Crystal or Plasma displays with lesser power requirement too.

Sony is the first manufacturer who plans to introduce FED-based displays in late 2009.

5.2.9 Surface-conduction Electron-emitter Display (SED)

The SED is a flat panel display conceptually similar to the FED. Developed by Canon, who began researching since 1986, SED provides the picture quality of a CRT on a flat panel with about 50% and 33% power saving respect to CRT and plasma displays, respectively having the same physical dimensions.

The SED consists of a series of electron emitters and a layer of phosphor, separated by a vacuum space. Every single pixel has a beam of electrons which does not require focusing and operates at a much lower voltage than a CRT, yet the brightness and contrast can be compared favorably with the high-end CRT (Fig. 28).

Recently (2004) Toshiba and Canon announced a joint development agreement to launch SED TV sets by the end of 2005, a goal which was also not met because of some contractual troubles so delaying mass production. Until today the SED technology is under development. However Samsung Electronics has shown an interest in SED technology as a replacement for current TV technology. Thus, it is not surprising that it would also pursue SED technology when it is made available.

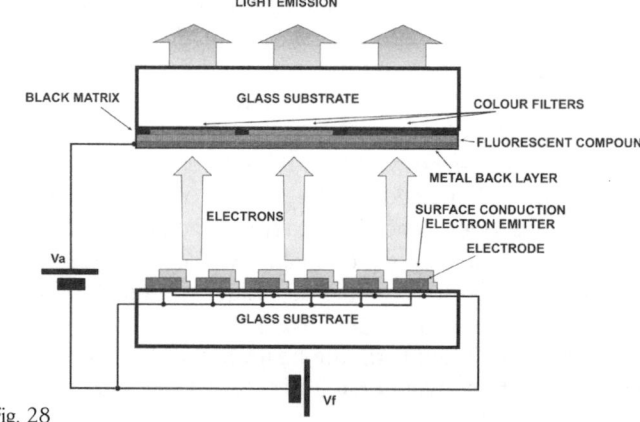

Fig. 28

5.2.10 The Liquid Crystal Display (LCD)

A colour display based on a completely different technology competing with CRT is the *LCD*, an acronym which stands for *Liquid Crystal Display*.

Liquid crystals were discovered in 1888 by the Austrian botanist *Friedrich Richard Reinitzer* in studying the liquid crystalline nature of cholesterol extracted from carrots. For the explanation of their behaviour he collaborated with the physicist *Otto Lehmann* but the substances could not be used commercially for displays until the 1960s because they were unstable at room temperature.

It was in 1962 that the researcher *Richard Williams* discovered the way in which light passes through a liquid crystal and changing its state of matter when it is stimulated by an electrical charge. On the basis of that discovery, in 1968 *George Harry Heilmeier* built the first prototype of a liquid crystal display.

LCD TV's popularity has recently exploded yet the LCD concept is at least 120 years old!

Liquid crystals are composed of *nematic* molecules in a state of matter close to liquid. Their physical behaviour is orderly but changeable whereby electrical stimulation caused by an electric field intervenes. In fact, the liquid crystals in a state of rest are placed in the cells in accordance with the helical structure so that light enters horizontally and twists vertically by rotating at 90°. If we put the liquid crystal in a layer sandwiched between two layers of polarizers, we can control whether the pixel lights up or not. A polarizer allows only light with a particular electric field orientation to pass through it. Only in this way the light can pass through the polarizing filter for vertical planes to finally excite the eye's retina.

It is evident in the case of the liquid crystal display that a backlight is needed and the image is created through the controlling of each LCD pixel using electricity. The liquid crystals could be roughly compared to a venetian blind with horizontal slats which can be rotated in unison from 0° to be parallel to sunlight for the maximum illumination to 90° completely blocking sunlight thus causing a quasi total darkness.

The colours are obtained by controlling the electric field in each colour LCD pixel made up of a red, green and blue cell.

The LCD TV or monitor can have a passive or an active matrix cell disposition.

The former (called also *STN=Super Twist Nematic, DSTN=Double Layer Supertwist Nematic, CSTN=Color Super Twist Nematic* and other variants) are usually composed of a grid of cells organized in rows and columns. To address a single pixel the lines are activated in sequence; this creates a slower addressing and is therefore less suitable for fast changes of content as required for moving images of footages.

Another flaw is the crosstalk. The cells activated by electrical impulse crosstalk with the electric field that activates adjacent cells, which adversely affects the definition of the image introducing luminous smudges and halos.

The colours are not too bright, the contrasts are faded and the angle of vision is practically fixed forcing us to be directly in front to watch the monitor (i.e. a narrow viewing angle).

Dual Scan technology has attempted to remedy these limitations due to the presence of two independent yet coordinated electrical circuits that energize the cells by reducing the time in which the liquid crystal does not receive an information bit. The image aspect is improved but the problem is not resolved, and this is the basis of low costs of a passive matrix display.

5.2.11 The Active Matrix Liquid Crystal Display (AMLCD)

The term was first used in 1975 by Dr. *T. Peter Brody* to describe a method of switching individual elements of a flat panel display by using a three terminal tiny transistor for each pixel called the *TFT* (*Thin Film Transistor*), used to hold the voltage of the cell until the next regeneration of a video image.

Each transistor is made of a thin and transparent film placed on the back layer panel that contains the liquid crystal cells resulting in a complete isolation of each pixel from its neigh-

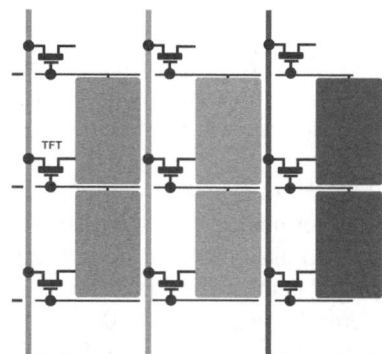

Fig. 29

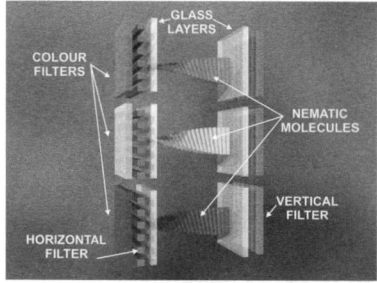

Fig. 30

bouring pixel (Fig. 29). The liquid crystal molecules do not tend anymore to return gradually to the state of rest before the activation of the next electrical pulse, as happens in a passive matrix display, nor should they maintain their position in the absence of voltage. Therefore the cells are faster to respond and the eye does not perceive flicker or light smearing effects and halos caused by crosstalk. Usually a storage capacitor is embedded in each pixel to improve display uniformity by reducing the pixel voltage offset.

This provides superior image quality. The colours are brighter, the contrasts are sharp and the viewing angle is not strongly restricted since it can reach up to 170° in height and laterally (Fig. 30).

Nowadays LCD technology is an amazing experience. High-performance TFT LCDs replaced successfully CRTs in applications ranging from airplane cockpit or car navigation displays to standard and high-definition video becoming rapidly the predominant technology for portable display applications such as laptops, PDAs or mobile phones. They offer a unique combination of low voltage supply, low-power consumption, low cost and optimal image quality.

5.2.12 HDR-TV Display

BrightSide Technologies, a team of researchers spun off from the University of British Columbia, have developed the world's first true <u>H</u>igh <u>D</u>ynamic <u>R</u>ange (HDR) display which uses LCD technology where the fluorescent backlight is replaced by several LEDs. Instead, a technology they call *IMLED* (<u>I</u>ndividually <u>M</u>odulated <u>A</u>rray of <u>LED</u> backlights) providing a luminance ranging from 0 to 3,000 candela/m^2 and a contrast ratio in excess of 200,000:1. This HDR display is the first capable of accurately displaying 16 bits per colour channel images.

Brightside has presently introduced a 37-inch display designed to satisfy the requirements of these 'vision critical' markets. More information about HDR displays is available at www.brightsidetech.com.

5.2.13 The Plasma Display Panel (PDP)

<u>P</u>lasma <u>D</u>isplay <u>P</u>anels (PDPs) were first invented in 1964 at the *University of Illinois at Urbana-Champaign* by professors *Donald L. Bitzer, H. Gene Slottow* and a graduate student, *Robert Willson* for the *PLATO* internal computer system. The first screens were monochromatic and appeared in either shades of green or yellow. The PDP technology was demised in the late 1970s because the classic CRT was cheaper to design and manufacture. Most people think of it as an innovative and state-of-the-art device, so you might be surprised to know that plasma TV technology is more than 40 years old!

PDP uses a mixture of two noble gases, *neon* and *xenon*, contained inside thousands of tiny

5.2 Display devices

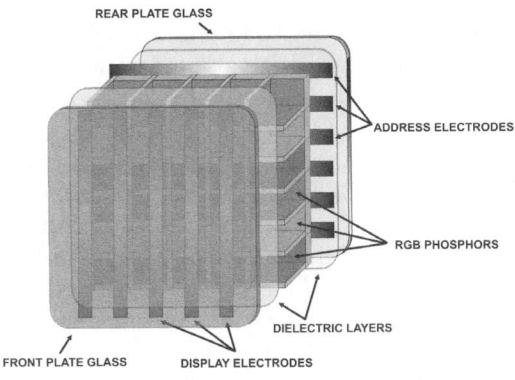

Fig. 31

cells positioned between two glass panels with a separation gap of about 100-200 μm. Two sets of electrodes are also inserted between the panels with the vertical address electrodes behind the cells and the transparent electrodes placed above the cells horizontally, forming an electrodes matrix.

In accordance with a video signal a dedicated microprocessor scans electrodes thousands of times in a fraction of a second. When the intersecting electrodes are charged, the electric current flows through the gas in the cell which, by stimulating the inherent atoms, is turned into a plasma (hence the name) during the process thus releasing ultraviolet (UV) rays.

Every pixel in a PDP is made up of three sub pixel cells with red, blue and green phosphors. Emitted UV rays interact with the phosphor at various levels to produce a wide variety of colours along the entire visible spectrum (Fig. 31).

Also, because each of the pixels in a plasma display is individually lit, the image is very bright and can be seen at almost every viewing angle.

Plasma screens do also have some drawbacks. The 'burn in' effect can occur when viewing a static display for a long period of time would cause that image to be permanently burnt into the screen. Plasma displays also produce much better blacks than LCDs and a good brightness compared to LCDs or CRTs. However it requires a higher and more precise controlling voltage thus consuming about four times the power over an equivalent CRT display.

Current plasma displays range in sizes as low as 21-inches to the gargantuan revealed at the *2006 Consumer Electronics Show* in Las Vegas measuring an amazing 103-inches, developed by Matsushita. Also with this technology being so new and despite prices that have been dropping in recent years, plasma screens are still relatively expensive in the display market.

5.2.14 Organic Light Emitting Diode (OLED) Display

OLED, also known as *LEP* (*Light Emission Polymers*) or *OEL* (*Organic Electro Luminescence*), is a display technology based on organic components (usually a carbon-based polymer) that emits light when current passes through it. Unlike LCD, therefore it requires no backlight. When the voltage is applied across the OLED, the anode is positive relative to the cathode with the effect of passage of electrons from cathode to anode. The cathode releases electrons to the emission layer and the anode withdraws electrons from the conduction layer (or creates electron holes in the electron conduction layer). Subsequently the electron holes pass the emission layer and rejoin with the electrons. When an electron finds an electron hole, the electron fills the gap and returns energy in the form of a photon. The nice results include a lower energy consumption and the possibility of building very thin OLED displays (Fig. 32).

Experiments on organic materials that exhibit electroluminescence have been conducted since 1950s: it is another very old technology! The low conductivity of the materials used in these early experiments, however, produced a limited light output which prevented their commercial use. Significant progress was made in 1987 from professor *Chin Tang* of Eastman Kodak

Chapter 5. Television Display Systems

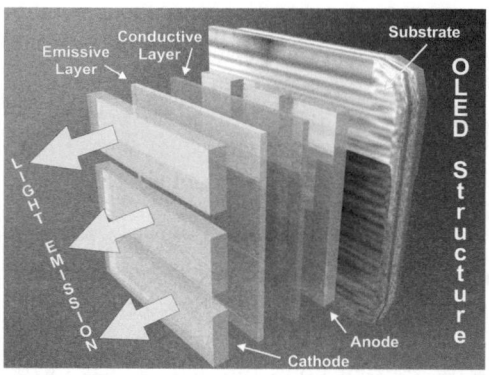

Fig. 32

who used a diode device to produce light emission from organic materials inserted between the anode and cathode layers. On this occasion was first used the term *OLED* and it became the basis for R&D of display technology based on organic material. In 1998, Kodak in collaboration with Sanyo exhibited the first coloured OLED display. The marketing of this technology began in the year 2000 when LG Electronics developed organic displays for its mobile equipment. This was followed by attempts to adopt the technology for high-definition displays. Thus came to market in November 2007 the first 11-inches OLED TV in the world from Sony with its *XEL-1* TV set.

In addition an OLED display can be printed on any suitable substrate with a very simple manufacturing process compared to LCD and plasma displays. Furthermore it produces benefits that match the best colour, black and brightness rendition because the light is produced directly from the OLED organic materials and the response time is faster than in the LCD display so being suitable for video revealing very fast movements.

The OLED displays also have a wide viewing angle, up to 160 degrees and require very little power, only 2-10 volts.

However OLED suffers some disadvantages. The main of these is the limited lifetime of organic materials compared to other display technologies like LCD and plasma. However, this problem has been addressed by the manufacturers for the development of techniques such that the OLED devices more efficiently return light from the screen. In addition water can easily damage the OLED display.

An OLED display seems a sci-fi dream that becomes a reality. Imagine a very large high-definition TV screen at least 1/2 cm thick or even a TV monitor that can be rolled up and can be placed in a bag after use or a screen device sewn into the sleeves of one's jacket!

Or imagine a future in which the newspaper and mobile TV are integrated into one device where the news on the devices are continually updating while we are moving! It is also more convenient to bend the device and put it in the bag after use!

Even nowadays this technology is actually used in a variety of equipment such as mobile phones, PDAs, MP3 players, car radios, digital cameras, and only recently, OLED display TV. Surely OLED is the most aggressive technology versus the most diffused LCD and plasma display technology which could be superseded as OLED and its numerous variants will be mass-produced at low cost.

5.2.15 Thick-Film Dielectric Electroluminescent (TDEL) Display

iFire Technology Inc. manufactures flat screen televisions that do not include vacuum, gases or liquids, making them more robust than other flat panel technologies with reduced sensitivity to shock, vibration and breakage. iFire displays emit light from the front of the display using a thick dielectric/thin-film phosphor device structure, with a consequent increasing in video performance as well as wide viewing angle and crisp, clear and colour saturated pictures. iFire flat panel displays boast a widest range of operating temperature of any electronic display technology, meeting the transport needs of other screens and high-reliability applications.

The company was founded in 1991 and is based in Toronto, Canada. iFire Technology operates as a subsidiary of Westaim Corporation.

iFire expects to commercialize its technology in partnership with major industry players, and plans to target the 30 to 50-inch screen size television segment in partnership with major consumer electronics companies. More information is on www.westaim.com.

5.2.16 WOWvx Display

Philips has developed a suite of technologies codenamed *WOWvx* that promises to help professionals in creating exciting three-dimensional (3D) viewing experiences.

The vision of a three-dimensional image is based on the fact that left and right eye see separated images. Therefore for the stereo vision of an artificial image it is necessary to differentiate the image intended for the right eye from that of the left eye and this is usually accomplished with viewing glasses having colour filters and coloured images matching those colours or polarized images and glasses with polarizer filters or alternating images in conjunction with synchronised shutter-type glasses, alternately obscuring and revealing the vision of the left and right.

However WOWvx technology provides 3D displays without the use of a specific lens. This is performed through a series of microscopic prismatic lenses in front of the pixels so that their colours are more focused towards right or left direction. However the lenses are transparent and therefore much of the colour passes through but the two eyes would receive a higher percentage amount of information from the appropriate pixels, or in other words, for each displayed pixel is assigned not only the two positions on the display but also a third fixed location along the axis perpendicular to the display, which provides a 3D depth effect.

The interface to WOWvx monitors relies on visualization of bi-dimensional (2D) images with the addition of information relating to the Z-axis (depth). Alongside the traditional 2D images is added a Z-depth image map (called the *Depth Map*) of the same size and resolution of the 2D original image. Each pixel of the depth map corresponds to a pixel of the 2D image and indicates the distance of the 2D pixel from the point of observation.

The 3D images are generated by means of creation tools (plug-ins, add-ons, etc.) to support the creation of 3D content from 3D animations or games, converting stereoscopic video content, or even upgrading existing 2D content to 3D. It is marketed by Philips as 'well positioned

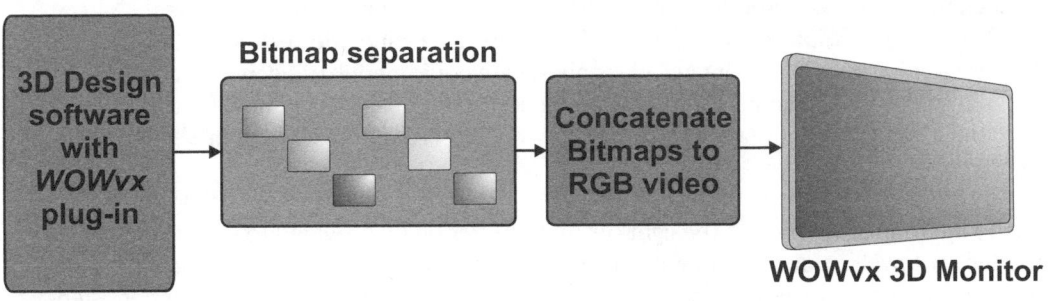

Fig. 33

for use in a wide range of consumer and professional applications such as digital signage, gaming, simulation and video' (Fig. 33).

The result is quite interesting but it is far from a true holographic 3D display.

5.2.17 Virtual Retinal Display (VRD)

This quite interesting display was invented by *Kazuo Yoshinaka* of *Nippon Electric Co.* in 1986 but in 1991 the *Human Interface Technology* Lab at the *University of Washington* developed a similar device. The significant difference from other display technologies is that VRD draws a line-by-line raster directly onto the eye's retina. Initially developed for military use, VRD spread also into the medical field and virtual reality applications.

Since the projected image must be focussed onto the retina, VRD is based on coherent ray lights created by a LASER beam which could became too hazardous to vision without severe and rigorous safety controls. Recently the researches are focussing on a safer and cheaper LED technology (Fig. 34).

5.2.18 Large screen displays

One method to visualize video moving images onto large screens is to somehow project them using magnifying optical media on a white cinema screen. This was achieved by a unit called the *Video Projector* that is essentially constituted by an equipment embedded with different types of devices driven by a very bright light to generally project the images on a specific screen.

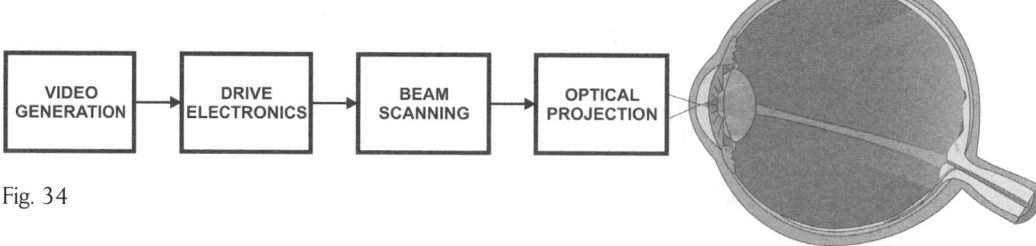

Fig. 34

A video projector quite similar to a motion picture film projector was the *Eidophor*, invented at the *Swiss Federal Institute of Technology* and by *Fritz Fischer* and *Edgar Gretener* in 1943 in Zurich but the original idea was conceived since 1939. The arc lamps were practically identical and projection lenses were used to magnify the desired images on the theatre screen.

Through the joint efforts of this Swiss group, CBS, and the technical staff of *Twentieth Century-Fox Corporation* the original black and white equipment had been converted in order to provide colour transmissions for theatre projection substantially using the CBS television colour sequential system, i.e. a rotating three-colour wheel inserted between the projecting beam and the screen.

In the case of the motion picture projector, a band of film carrying a series of still images passes intermittently through the light beam, a shutter being provided to cut off the light while the film is travelling. In the Eidophor projector, instead of photographic images on a strip of film, a sequence of images is created on a thin layer of a special liquid (similar for consistency to honey) which is placed on a slowly rotating mirror surface in a position optically equivalent to

5.2 Display devices

Fig. 35 - Eidophor projectors in use at Nasa's Mission Operation Control Room

the position of the film in an ordinary projector. The succession of images on the thin liquid layer was produced by means of electrons deposited on the surface of the liquid. These electrical charges are proportional to and controlled by the television signal, in much the same manner that a television signal is used to produce an image on a regular TV picture tube.

The essential difference is, of course, that in the home receiver tube the electron beam strikes the end of the tube and causes the phosphor material to glow point-by-point and line-by-line, with a brightness which is proportional to the point-to-point brightness of the original scene.

In the Eidophor projector, however, the electron gun caused the liquid to take on tiny surface irregularities and thus to change its optical properties. The picture thus produced appeared very much like the relief image in hardened gelatine used in some photographic processes, after the silver image had been bleached away. By means of auxiliary lenses and properly arranged mirrors the instantaneous picture on this 'image bearing' layer of liquid was finally projected to the viewing screen.

The Eidophor was eighty times brighter than CRT projectors of that time and produced high quality images up to 18 meters in width. The Eidophor was a large and cumbersome device of excellent image quality. It was used by the NASA space program, where the technology has been deployed in mission control (Fig. 35).

The production was completely stopped in the late 1990s when technological evolution produced smaller and more efficient *CRT, LCD* and *DLP Projectors*.

LCD Projectors are based on the filtering of a very bright lamp light by arrays of liquid crystal displays. This is the simplest system, making it one of the most common and affordable for commercial and personal use. Its most common problem is a visible pixel effect, although recent improvements have minimized this.

DLP (Digital Light Processing) Projectors are based on modulation of the reflection of light by using a *Digital Micromirror Device* (DMD), a device invented by Texas Instruments in 1987 (Fig. 36).

Fig. 36 - Picture of DLP chip used in a digital projector at the Cinerama in Seattle. Photo by *Andrew Hitchcock*

Three-Tube CRT Projectors are reliant on technology of traditional cathode ray tube televisions. These last are the oldest system still in regular use, but falling in favour of LCD projectors due to their unwieldy cabinet. However, it provides the largest screen size for a relative low cost.

Basically a video projector is able to render very large images even on a wall or on a theatre screen sourcing images from a computer, a

151

Chapter 5. Television Display Systems

Fig. 37 - A Zenith 1200 CRT Projector based home theatre. Around 2006

TV tuner, camcorder or any video source. The projector is usually fixed under the ceiling of a room, at a projection distance of about three-four meters or more, but may be positioned in other wiser locations (Fig. 37).

The bright light source is generally one or more halogen or <u>U</u>ltra <u>H</u>igh <u>P</u>erformance lamp (or <u>U</u>ltra <u>H</u>igh <u>P</u>ressure, shortly *UHP*, usually mercury arc, originally developed by Philips in 1995). Just recently some brands (Sony, Mitsubishi, 3M, etc.) have introduced high brightness LED technology for projection lamps which in conjunction with a LCD display device can realize low-cost lightweight pocket video projectors even supplied by internal batteries!

The video projector is also employed at home in combination with an audio *Home Theatre System*. In public, the projector is used to show videos and documentaries to a wide audience, in universities and schools to project notes and lessons to students, allowing them greater learning and removing the old blackboards on which many times the writing was incomprehensible to the student.

Recently, video projectors are used very well in outdoor live concerts or similar performances to show video clips to public masses while the band sings or the artist performs. In addition the projector can be placed behind the screen for back-projecting the video source. In this case a proper 'electronic mirror' function key flips the image horizontally and the projection screen has to be semitransparent.

In 1970 the first *Laser Video Projector* was designed. Lasers may become an ideal replacement for UHP lamps, which are currently in use in projection display devices such as rear projection TV and front projectors. Since current televisions are capable of displaying only half of the visible spectrum of colours, one of the main claims of lasers is the ability to produce undiluted, perfect shades of colours that allows precise mixing. With the colour enhancement capable with lasers, up to 90% of the currently non-viewable spectrum could be recovered.

Designers discovered the use of special mirrors for the simultaneous lasing of green, red, yellow, and blue wavelength beams can be seen by the eyes as white. A rotating polygonal and a curved mirror allow the laser beam for horizontal and vertical refresh modulation respectively to project video on any type of surface projection. Contrast, colour space and sharpening are higher than that of other projection technologies.

Nowadays solid-state laser generators, such as laser diodes, provide laser video projectors with the highest luminous flux output and a richer more vibrant colour palette than the conventional DLP, LCD or CRT projectors.

Arisawa, Epson, JVC, Philips and DLP (Texas Instruments) have established the *Micro Device Display Consortium (MDDPC)* in order to widely publicize the features and advantages of *Micro Device Display Projection Televisions (MDDP)* among consumers, retailers, industry experts and the mass media. The new consortium is aimed at promoting the growth of MDDPs and boosting public awareness of the Technology.

This Consortium wants to produce rear projection systems under the name *MDDP (Micro*

5.2 Display devices

Device Display) which consist of very small pixel composition (several micrometers) using semiconductor processing technology. There are several types of devices such as micro mirror devices using very small mirrors on silicon, *LCOS* (*Liquid Crystal On Silicon*) using liquid crystal silicon and reflective contacts and *HTPS* (*High-Temperature Polysilicon*) using liquid crystal silicon with transmissive technology. The MDDP is a projection display that expands an image from this device to the screen. The current size of Micro Display devices is around 0.5 to 0.7 inch in diagonal.

MDDP TV sets first began to appear in the market around 2004. MDDP TVs outperforms conventional CRT projection televisions in brightness, high resolution, wide viewing angle, light weight and superior design by a wide margin. As MDDP TV sets offer high value cost performance based on their high image quality and large screens, they have gained a particularly strong following in North America. More info on www.md-display.com.

5.2.19 Multiple screen displays

In order to obtain very giant and custom flexibility screens, multiple unique displays can be assembled to form a single tiled large screen. The display technology behind this is therefore the same of that a single device, as CRT or LCD for example. The simplest one is the dual monitor display (commonly known as *Dual Head*, even if some cards furnish a *Triple Head* output and even more!) provided in earlier times by professional computer video cards only and today by almost any consumer card, along with a supported video driver under regular operating systems. Generally onto the primary monitor the graphic designer visualizes the work area while at the same time he watches the results on the secondary one.

On video fields such multiple screens are called a *Videowall* and start from a 2x2 display matrix to ideally without any superior limit. Videowalls generally respect the aspect ratio of a single 'cell' of the matrix, i.e. a single picture display. So if for example we will build a 4x 4:3 videowall screen, it will be composed by 4 rows x 4 columns = 16 4:3 single screens or more proportionally multiplied by, even if it is not so strictly binding.

Each of the available screens is configured to only show the section of a video frame that corresponds to its position within the array of screens. Therefore, a digital video frame storage is needed to split the video material to be shown into different sections. More sophisticated video storages have the ability to accept external multiple video sources to be displayed on single cells, alongside with the main source.

Fig. 38

Videowalls are widely used in television indoor studio shows, outdoor live concerts, etc., but own a severe drawback. Since the screens are tiled and completely separated one from another, some narrow 'black lines' can appear on the videowall caused by the screen's frameworks (Fig. 38).

To counteract this problem a new generation of videowalls was developed relying on LED technology by using LED panels in place of traditional CRT or LCD screens.

The peculiarities of LED screens are the

Chapter 5. Television Display Systems

high brightness, the resistance to weather and the modularity, i.e. the possibility of reaching any kind of size. The *LED Videowall* (or simply *LEDwall*) can be installed outdoors, in full daylight, hung or placed at any height.

Available in a wide variety of pixel pitches and LED configurations, high light output, superb image quality, long-term stability and elevated lifetime make LEDwall a great choice for applications such as exhibitions, TV shows, concerts, out-of-home advertising, sporting events and public large assemblies.

Unlike the alternatives, LEDwall can deliver images with real punch in high ambient light and does not suffer from screen burn and is not reliant on expensive and sensitive light sources. In addition the contrast levels in outdoor environment are undiminished by full sunlight and the uniformity of colour from corner to corner is practically perfect.

However this high quality, performance, flexibility and reliability have a cost. A LEDwall is actually priced at several thousands of Euros, being composed by panels each one at least with 64x48 pixels at full RGB colours (every pixel is made of a single RGB LED or a cluster of three single primary colour LEDs as shown in Fig. 39).

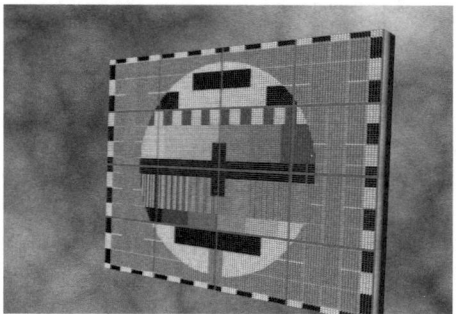

Fig. 39 - A 3D rendered LEDwall showing a PM5544 monoscope

The principle of functioning is the same of the other television technologies, i.e. the image exploration line-by-line and frame after frame, with the evident difference that here every single LED pixel can be individually addressed by a suitable computer system. In addition the LED brightness modulation can be finely adjusted using a *PWM* (*Pulse Wave Modulation*) technique.

The idea to use LEDs for arranging a screen is old, however. Since LED's invention several LEDs were grouped to form up visual panels for displaying messages, advertising or informations. This was one of the main professional use, limited in early days only by the high cost and reduced choice of colours, i.e. red, green, yellow or orange utilized as single colours (monochrome LED display) or in combination (bicolour, red-green or red-yellow, etc.). The first blue LEDs were made in 1971 but they had too little light output to be of much practical use and the fabrication process was complex and expensive. Advances in LED technology and doping techniques in recent years brought blue LED increasingly being used in applications more traditionally associated with consumer market with consequent reduction in costs.

Nowadays blue LEDs are widely diffused at a relative low cost for small quantities to cover in combination with red and green LEDs a full light visible spectrum and even over with infrared (IR) and ultraviolet (UV) LEDs.

5.3 Things To Think About

Picture tubes are one of the last display devices that we have to deal within the television world because nowadays they are cheap and do the job well. There are other new technologies available and the user might fear buying an alternative display device that could become

5.3 Things To Think About

obsolete within a few years with big repair issues because of a lack of spare parts or the repairing cost exceeds that of the same apparatus!

Today's broadcast facility is almost guaranteed to contain both analogue and digital signals. We often forget that reference signals today are still based upon NTSC or PAL characteristics. During the transition from analogue to digital, i.e. from standard displays to real fully digital high definition facilities, we need to embrace tools that help us bridge multiple worlds.

Just like the transition from B&W to colour television we have not suddenly thrown out the old TV set with the rubbish to acquire a new colour set, for the same reason the changeover from completely analogue to completely digital televisions will happen gradually. This trend is confirmed by simply observing the numerous analogue inputs present in a new generation television set, despite of one or two *HDMI* (*H*igh-*D*efinition *M*ultimedia *I*nterface) inputs, historically (2002) the first connection standard able to carry both audio and video uncompressed digital signals.

Returning back to video display devices, it is difficult now to hypothesize what technology will be the ruling or winning.

Every new introduction has pros and cons. CRT devices for example have a consolidated service network and is highly supported by spare part producers but are voluminous, heavy and, under a certain profile, dangerous to handle or repair. Therefore slowly it will be discharged and replaced.

On the other hand, liquid crystals inside the LCD display are poisonous and must not be ingested or brought into contact with skin but are efficient, energy-saving and light. Plasma display are too power hungry and rumours say it will be banned for this reason. OLED seems to have the most chance to become the incumbent technology and actually it is under a quite fast development phase. We suppose that OLED mass production will explode in popularity within a decade.

Digital display technology gave and still gives a great boost in image displaying even through interactive screens such as touch and multi-touch screens. However it is very distantly into the future before we will see a real 3D holographic display such as that seen in science fiction movies such as John Woo's *Paycheck*. Or the computer projected images in Steven Spielberg's *Minority Report* movie. Or even the full interactive 3D simulated reality facility called *Holodeck* of the *Star Trek: The Next Generation* TV serial! Nowadays these latter devices are only sci-fi display devices!

6. Video Cameras

In television the capture of moving images is achieved by special equipment commonly called a *Video Camera*, or simply a *Camera* which, roughly speaking, should imitate the human eye. The task of the camera is to convert the optical image of the subject framed by some optical means into electrical signals via some light photosensitive devices known as *Pickup Devices*.

The camera is the most creative and exciting part of this long electronic transformation process, from a television indoor studio or outdoor places to our houses. It is responsible for the acquisition in almost real time of the reality that surrounds us, making such information available to the deployment process. The camera then deals with the real world, it must produce images of real, not synthetic, things and must confront the outside world in which it must work, capturing reality and its infinite forms of representation. It must often work in hostile environments despite its extreme constructive complexity and must be practical and above all extremely reliable whilst providing the highest quality possible.

Fully electronic cameras can be divided into two main families based on two very different technologies. The earliest huge electronic cameras used since the middle of 1940s until around the end of 1990s were based on one, two, three or even four electronic pickup tubes. Afterwards a solid state component based on CMOS or CCD technology permitted the design of smaller high-quality colour cameras at lower costs, which utterly invaded the camera market. Tube-based cameras have, nowadays, almost disappeared in television facilities, being completely replaced by CCD-based cameras. However pickup camera tubes are still used in some non-broadcast, military, niche-medical and specialized sectors.

As widely mentioned previously, Baird's mechanical cameras based upon Nipkow disk were the first pickup devices available in the early days of television, but their resolution was limited to 240 lines of resolution by use of very noisy Nipkow disks of over 1 meter of diameter which revolved at 50 revolutions per seconds with all the vibration that brought! Though mechanical cameras could never scan and deliver clear, live-action and fine-detailed images, most would-be TV inventors still hoped to perfect it! It was clear that the incoming mass-media called *Television* would be based on different and new camera technologies.

Moreover, television evolution has always been closely linked to technical progress of the cameras and, more particularly, to the progress correlated to capture devices, i.e. the primary items that come into play for the creation of an electronic image. Since this is a starting point, there follows a description of the process of formation of the electronic image as well as the related elements that have been used.

6.1 Tube-based camera devices

The first experiments to take pictures using electronic devices were based on cathode ray tubes operating almost backwards. The electron beam's target was a photosensitive area which changed its electrical characteristics when struck by focussed light from the image of the framed object, rather than emitting light when electrons hit phosfors as it happens in picture tubes.

The generated signal, amplified and properly processed, could be immediately transmitted over the air to your television or, if the technology could permit it, could be 'stored' directly in a video registration system.

6.1 Tube-based camera devices

Regardless of their specialized features, all camera tubes invented, developed and refined over the years were based on the same principles of construction, schematically divided into three sections: an electron gun, a deflection circuit and a target. Over time, the discovery of high-sensitivity materials with excellent light responses and great performances, essential for colour television, enabled the development of photoconductive targets. There the transformation of light radiation into electrical signals exploited the characteristic, typical of some semiconductor materials, of photo-dependant electrical resistance depending on the intensity of impinging light.

Below are listed the principal camera tubes invented from the dawning of electronic television until the first solid-state cameras, used nowadays.

6.1.1 The first experimental camera tubes

The turning-point in the scanning system was the invention of the first video camera tube (known as an *Image Dissector* and built by *Philo Farnsworth*) bringing a significant improvement in the quality of the images hitherto generated mechanically and thus paving the way for a fully electronic television. His invention in 1927 was a practical implementation of a theoretical system described in a 1925 patent from the German inventors *Max Dieckmann* and *Rudolf Hell*. In 1934 he largely succeeded in demonstrating the potential of a fully-electronic television system having 220 lines and 30 frames per second, which he had developed from that first prototype.

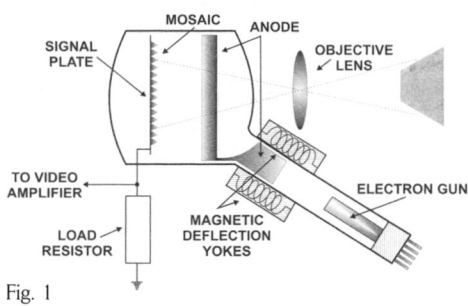

Fig. 1

That first device, however, had a very low light sensitivity and therefore required that target objects were strongly illuminated. For this reason it was technically outdated within a few years as a regular television capturing system, although still used for special needs of televising bright objects for example, metal castings in process.

The *Iconoscope* (Fig. 1), developed by *Vladimir Zworykin* in 1932, was probably inspired by Farnsworth's dissector and brought a significant improvement in the quality of images generated electronically over the invention of his rival.

It essentially consisted of a cathode ray tube having a photoelectric mosaic laid on the front screen, thereby functioning similar to receptors of a human visual system. At the rear side was placed an electron gun constructed so to emit electrons when heated by a filament called a *Cathode*, and a series of electrodes (grids) that accelerated and focused the electrons to form a very narrow and fairly compact beam, all enclosed in a glass vacuum bulb.

The photoelectric mosaic was constructed by a metal plate, called the *Signal Plate*, covered with microscopic photosensitive cells, isolated both from the plate and between them. Each microcell then submitted some electrical capacity respect to the signal plate to form up a kind of very tiny capacitor.

Construction of the photoelectric mosaic started with a thin sheet of mica having one full metallic face and the other side covered with a very thin layer of silver deposited under vacuum by an evaporation process. Subjecting the microscopic grains comprising silver film to a treatment of oxidation in caesium vapours, created a caesium oxide coating. By this treatment each granule became a tiny photocathode with spectral sensitivity extending right across the visible light range of frequencies.

The metallization of a ring-like area of the tube's bulb, facing the photoelectric mosaic, constituted the common collector anode to all the photoelectric cells. In the absence of incident luminous radiation, all the elements of the photoelectric mosaic and their respective electrical capacitances had the same potential as the signal plate and therefore the respective image capacitors were in discharged state with respect to the signal plate.

When an image was focused onto the mosaic, the tiny mosaic's photocathodes emitted electrons towards the anode in proportion to the intensity of the light striking them. Thus, the photocathodes acquired a corresponding electronic image of positive potentials compared to the plate signal, corresponding to the distribution of brightness in the optical image. Each mosaic's photocell, being isolated from all others, kept its positive charge in the respective 'image' capacitor, until discharged was accomplished by the action of the negatively charge of the electron beam being scanned over it.

The electron beam explored the entire surface of the mosaic in parallel lines, under the command of horizontal and vertical deflection yokes driven by a sync pulse generator, just as in a kinescope, thereby transforming the electronic image of the mosaic into an orderly sequence of electric pulses, thus providing a video output available at a signal plate load resistor.

Essentially the electronic beam worked as a switch for each element of the mosaic (represented as a photocell in conjunction with the common anode), and with a capacitor corresponding to its capacitance with respect to the signal plate. The electronic beam acted as a separate circuit discharging the capacitor, thereby providing a pulse of current through the load resistor.

In practice, the process of converting an optical image into an ordered series of electrical pulses was less straightforward than described mainly because of secondary emission (i.e. slow electrons from an area hit by fast electrons) produced on the photoelectric mosaic by the scanning electron beam.

The mosaic had a strong power multiplier which, near the surface, caused an appreciable electron cloud which meant that the all the elements of the mosaic did not behave in the same way at the same illumination.

Indeed, because of the presence of secondary electrons between neighbouring photosensitive elements, it could raise electronic currents which tended to destroy the differences of potential between adjacent elements, with negative consequences on the sensitivity and fidelity of the iconoscope. Another drawback was the unusual angle that the electron gun faced in relation to both the optical axis of the tube with a perpendicular plane of the mosaic. This inclination imposed the adoption of specific measures to reduce the size of resultant trapezoidal distortion and to keep in focus the electron beam over all the points of the explored surface.

As a pickup tube for television cameras the iconoscope also had many other disadvantages. The most important of these was perhaps the need for relatively powerful light sources to generate video signals of acceptable quality. The iconoscope was almost unusable in sub-optimal illumination but it was a valid image-to-electrical transducer under controlled light conditions.

6.1.2 Further developments on camera tubes

The next development of the camera tube was an improved iconoscope (called the *Image Iconoscope*) and later (1937) a family of tubes called *Orthicons, Multiplier Orthicons* and *CPS Emitrons*. These tubes were very similar to the iconoscope but the mosaic-target was now placed on an insulating membrane whilst the signal plate on the opposite side was transparent. The name

6.1 Tube-based camera devices

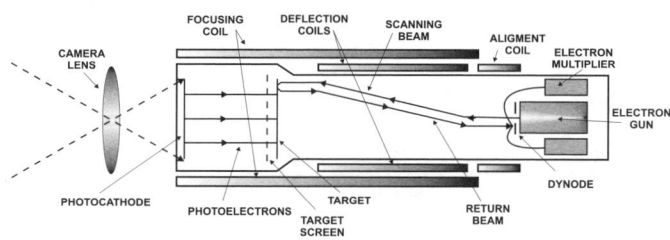

Fig. 2

Orthicon derives from the fact that the electron beam bombarded the target orthogonally at low scanning speed, while the CPS Emitron takes its name from the process of *Cathode Potential Stabilized* target scansion.

In RCA laboratories between 1939 and 1940 alternative models of camera tubes were created which substituted the dissector, iconoscope, orthicon and emitron. Such tubes, known as *Image Orthicons* (Fig. 2), were largely employed both in the military sector and later in civil applications.

Inside the image orthicon, electrons released from the photosensitive surface emerged parallel to it 'redrawing' the image electrically in the adjacent space. Then those electrons were accelerated by an externally applied voltage towards a thin glass plate (about 2.5 microns). The electrons arrived at and hit the plate with such energy so as to expel several secondary emitted electrons. This created a positive ionic image on the plate (due to emission of these secondary electrons) which electrically 'duplicated' the real picture in positive charges.

At this point the plate was explored through a beam of electrons emitted by an electron gun. Apart from this last detail, the technique used here was totally different from that of the iconoscope. First, the scanning electron beam approached the plate, being slowed down by an electrostatic field, which then accelerated the same beam back, directing the electrons to a common anode (called a *dynode*). The different points of the plate were positively charged according to the brightness of the image: the plate would therefore 'steal' electrons from the electron beam according to the amount of positive charge at every single point. In this way the amount of electrons which would return to the dynode would be the difference between those emitted by the electron gun and those attracted from the plate to neutralize their positive charges. Based on this difference, which varied each time the electron beam moved from one cell to another one, it was possible to reconstruct the 'image-by-voltage' of the scene. This system, though more complicated, presented less overall technical problems than the iconoscope and therefore it was adopted for television broadcasts.

The image orthicon owned the hitherto unusual but desirable characteristic of being able to generate a suitable video signal in any lighting condition, from sunlight to candlelight, also thanks to an internal electron multiplier as well as its logarithmic curve of response to light being quite similar to that of the human eye. Despite these brilliant features this tube possessed some unsolved anomalies such as a dark halo around bright sources of illumination (e.g. a spotlight) or white objects strongly lit. The system of automatic gain control of the video amplifier tried to correct the unbalanced white by acting on the image contrast. What was then obtained was an image where a blinding white subject 'glowed fierily' in the display device, surrounded by the rest of the image now too dark, almost overexposed [1].

Image orthicons were the first camera tubes to be employed in assembling the earlier colour camera (based upon a revolving three-colour wheel filter) because their high sensitivity to the light counterbalanced the inefficiencies of the optical systems of those times.

Around 1950 in RCA laboratories a new camera tube was developed as a cheap and

Chapter 6. Video Cameras

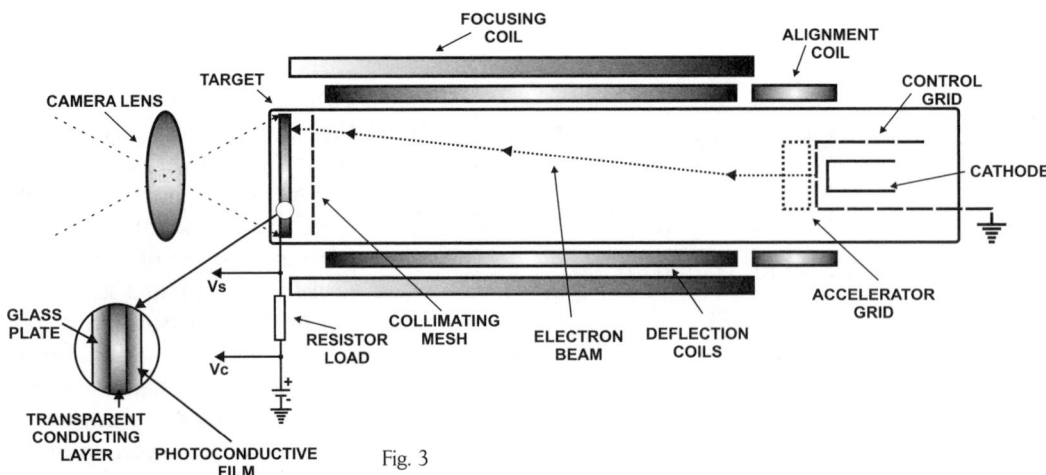

Fig. 3

simple alternative to the structurally and electrically complex image orthicon. This new device was called a *Vidicon* (Fig. 3). It employed a layer of cadmium sulphide and a photoconductor layer made of antimony trisulphide. Its dimensions, its lower cost and good sensitivity made it particularly suitable for portable industrial and amateur cameras, although it could not boast the same depth of features as the previous type of camera tubes.

After the vidicon tube came further innovations to produce better pickup tubes. Hitachi trademarked the *Saticon*, a tube used only as a professional device used in 3-tube broadcast and trichromatic monotube cameras. Its sensitive target was composed of *Selenium Arsenic Tellurium* (SeAsTe) from which its name was derived. It was also produced by Sony and Thomson.

Matsushita developed an interesting tube known as the *Newvicon*, whose target was made of *Zinc Selenide* (ZnSe) and *Zinc Cadmium Telluride* (ZnCdTe), two compounds very sensitive at very low illumination (night vision).

The *Trinicon* was a trichromatic monotube produced by Sony which used a special vertical colour stripe filter on the target. The Trinicon was used both in amateur and some semi-professional cameras.

The *Resistron* was developed by Heimann as a variant of the vidicon tube. Later, the *Pasecon* was developed, where the surface target was composed of *Cadmium Selenide* (CdSe). Toshiba manufactured the same type of tube, which it renamed the *Chalnicon*.

The *Silicon-Vidicon* camera tube (or *Silicon Diode Array*) was a rather unknown pickup tube, in which a matrix of about 300,000 silicon photodiodes was used to form a sensitive rectangular target, one photodiode for each pixel. The signal was extracted from each diode by electronic scanning as in the vidicon. Such tubes could be considered the link between a fully semiconductor device and an electronic tube. Application markets were industrial and educational sectors, before the CCD age.

Philips, in the early 1960s, produced a revolutionary pickup tube known as the *Plumbicon* (Fig. 4). It was also manufactured by other different industries under various names (*Vistacon*, *Leddicon* or *Hi-Sensicon*). It was used in most professional 3-tube studio and portable cameras. The target consisted of a transparent sheet (signal plate), made of tin or indium dioxide, a strong *N-type* semiconductor. Two layers of lead oxide were deposited on the scanning side of the target. The first of these two was almost pure lead oxide, a 'non-doped' (*Intrinsic*) semiconductor. The second layer of lead oxide was doped to form a *P-type* semiconductor. Altogether the, now, three layers formed a diode with a P-I-N junction. Due to these features, plumbicon tubes could

6.1 Tube-based camera devices

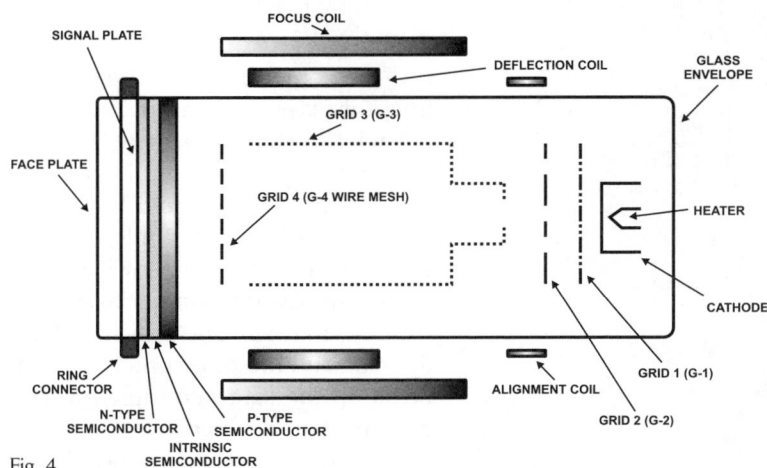

Fig. 4

be considered as a kind of 'semiconductor-tube' hybrid which, in comparison to other image tube technologies, offered high resolution, low 'comet tail effect' and superior image quality with very low noise and highly linear light response, but, unfortunately, all this at a very high cost. In fact each tube was priced (in 1960s) at least of 10,000 US dollars!

Plumbicon tubes are still produced by the American company *Narragansett Imaging* (in North Smithfield, Rhode Island in the United States) as spare parts (for some type of saticon tubes too) for old cameras still in operation or for television intensifier systems used in conjunction with medical and industrial x-ray units.

This list of pickup tubes will finish by introducing a sensor sensitive to infrared light (heat). In the early 1970s cameras equipped with such sensors had always been expensive and bulky equipment, and required complex cooling systems, and were therefore employed mainly by the military.

In the late 1970s and from the beginning of the 1980s, a new 'uncooled' technology was developed utilizing as thermal sensor a *Pyroelectric Vidicon* tube whose effect was to produce voltage by variation of temperature, obtained via crystals of *Triglycine Sulphate* (TGS).

Since the late 1980s, thermal imaging technology has been improved to attend to solid state sensors used extensively for earth exploration, radiation monitoring and astronomical telescopes. In the mid 1990s, it also made a first appearance in the automotive, rescue, industrial and security fields.

6.1.3 1, 2, 3 or even 4-tube Colour Cameras

We know that the process by which we obtain colour television is based on the decomposition of the image into three primary colours (red, green and blue) and their subsequent additive mixing.

Therefore colour cameras must have a system capable of breaking down the image framed by the objective lens into three independent images (red, green and blue), and generate from these three separate electrical signals, each one corresponding to a primary colour. This process can be achieved using different optical systems which will respectively feed three, two, one or even four pickup tubes.

The image orthicon was the first pickup tube to be used in the early colour television system and the last one for broadcast B&W television, although some low-end portable B&W and colour cameras were equipped with vidicon tubes. As electronics evolved and new tubes were invented, more efficiently colour separation methods were developed rather than simply three primary colour filters. One of these methods was (and still is) based on *Dichroic Mirrors* (Fig. 5) which are one of the most important optical elements in a colour camera.

Chapter 6. Video Cameras

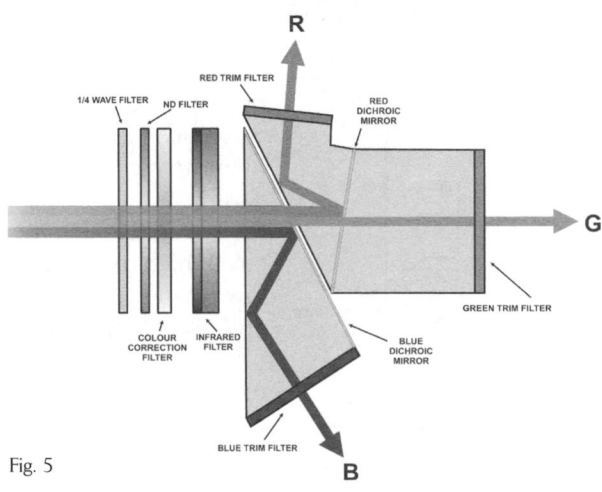

Fig. 5

These mirrors are made of multiple thin layers of transparent material, which posses alternately high and low indices of refraction. Depending on the thickness of these layers, the dichroic mirrors reflect only the light of a particular colour, allowing other colours to pass through normally. Two dichroic mirrors, one which reflects the blue and the other one the green, break up the light into three primary colours that, through normal reflective mirrors, are directed towards the targets of the corresponding pickup tubes. The light from the objective lens first strikes a dichroic mirror calibrated to reflect the blue, but the red and green pass through this mirror. The blue light is then directed through a mirror to the conventional pickup tube for capturing this colour.

The green and red colours, which have passed through the first dichroic mirror, reach the second mirror, designed to reflect red light; the green passes through to the second pickup tube. The red light, finally, is sent through a second mirror to the third conventional pickup tube.

In this way, each tube provides a video signal corresponding to that primary colour to which it is excited while the three signals (red, green and blue) are collected and sent to a mixing circuit. As widely explained previously, the luminance signal is obtained by mixing the colour signals coming from each tube, according to a fixed proportion of 30% for red signal, 59% for green signal and 11% for blue signal.

From this point onwards the colour encoding process begins following all the techniques and artifices typical of the television standard (PAL, SECAM or NTSC) for which the camera was designed, bought and will be used. The encoding system has been exhaustively discussed previously. Here we can add only that 3-tube cameras are used by television broadcasters, both in the studio version of productions in place (e.g. shows, news, etc.), and in standalone lightweight cameras used for reporting and outdoor shoots in general (the so-called _Electronic News Gathering_, shortly ENG, or _Electronic Field Production_, shortly EFP). In the first case, the cameras provide only the generation of video signals, while the processing, coding and sync generation are outsourced to an external unit called a _Control Camera Unit_ (CCU) which coordinates and controls all the cameras that are working in the television studio. Professional ENG cameras, however, are autonomous since they generate the colour video signal aided by an internal SPG and a colour encoder. They also offer the possibility of locking in sync with other cameras for studio mixing and recording, by use of an external synchronization system known as the _Genlock_, an important device which we will discuss later.

However, because the first 3-tube cameras were heavy and had a significant energy consumption, which made them generally unsuitable for outdoor use, some pickup camera manufacturers tried to design tubes with smaller dimensions and to develop pickup systems that required a lesser number of tubes.

2-tube cameras were used both in industrial and in amateur fields. For the colour decomposition it was used a single dichroic mirror designed to reflect the green light, which was collected by one of the two tubes. The blue and red lights, passing through the dichroic mirror, reached the second tube that had in its front a filter consisting of vertical stripes sensitive to blue light, arranged on a background layer sensitive to red light. With this type of filter, called a *Stripe Filter*, the signal plate collected, under the action of the electron beam, an electrical signal corresponding to the frequencies of the stripes sensitive to blue, from which via a decoding process would extract the colour component corresponding to red.

Once obtained, the three primary video signals proceeded, as previously mentioned, to their mixing and encoding according to the camera television standard.

While 3-tube cameras were commonly used in professional video environment and 2-tube systems had broadened the scope of utilization of colour cameras also in the industrial and amateur fields, the biggest innovation in camera technology, however, consisted of a new filtering and signal processing system that made it possible to use a single pickup tube. This system was (and still is) used in all amateur and much of the industrial and semi-professional cameras. Generally a vidicon, saticon or trinicon tube was used.

In the case of the saticon tube, there was a filter placed at the front end of the tube, i.e. in front of the target, that consisted of transparent and coloured bands of two types: cyan (blue + green) and yellow (red + green). These bands were the same size and overlapped. When the light affected the transparent areas, all three colour components could move toward the target, and when the light hit a coloured area of the filter, one of the primary colours was reflected. The result of this filtering, such as performed on the yellow band, adjusted the modulation of a carrier wave. Then this waveform was sent to the circuits that provided separate signals corresponding to blue and red colours. The third colour signal (green) was derived from the unmodulated component of the signal generated by the tube.

Incidentally Philips and Bosch designed and produced some models of studio cameras equipped with even four pickup tubes (in this case four plumbicons), one for each primary colour and the fourth tube for greyscale shades only (to separately detect the luminance Y), due to the registration difficulties of the early plumbicon scanning yokes. As yoke manufacturing improved, it became easier to realise the high registration accuracy necessary for 3-tube cameras, eliminating thus an extra expensive tube.

In addition, since the plumbicon's colour response could be varied by the manufacturer, plumbicon tubes with spectral responses were built for each of the primary colours, identifying each tube by the letter R (red), G (green), or B (blue) following the basic number. For this reason plumbicon-based colour cameras adopted prism mirrors instead of dichroic mirrors in order to obtain the best image quality.

However the first 3-plumbicon cameras were arranged in the form of YRB pickup tubes with the handicap that the Y channel did not give the same greyscale response to colours, resulting in a red or blue dominant colour different from one camera model to the next. Red or blue trim filters were placed in the optical path to equalize the colours. However with this expedient the Y or luminance channel gave a sharper picture with details in all colours. For cheapness later standard and improved RGB tubes were utilized.

For a television network, such studio cameras represent a significant investment (several thousand of dollars) but even in spite of their old technology they are still used in some special events where the smoothness of the generated images can be compared in the most part to the current broadcast solid-state modern cameras!

6.1.4 Pickup tube features

Pickup camera tubes have a number of structural features and qualities that ultimately determine the yield of the television image.

Regarding the pickup tube, a key feature is the diameter, measured in inches. The most frequently used sizes are one inch for each tube employed in colour professional studio and some portable cameras, with a target size of 9.5 x 12.5 mm, and inch for industrial and amateur cameras, with a target of 6.6 x 8.8 mm, although smaller tubes ($\frac{1}{2}$ inch) were produced for surveillance cameras or other uses. The earlier B&W pickup tubes (dissector, iconoscope, image orthicon, etc.) had sizes of 3 to 7 inches (Fig. 6)!

Fig. 6 - 4.5" and 3" monochrome orthicon tubes compared to a 1" colour vidicon tube

Another feature is the ability of the tube to analyze finer details and it depends mainly on the size of the particles that make up the target. The smaller they are, the higher is the definition of the tube and, consequently, the yield of the camera.

The *Definition*, more accurately called horizontal resolution, is expressed, as previously stated in the case of televisions, as number of lines per frame; in cameras for professional use it was (and still is) generally greater than 500 lines, while for amateur use it is about 300 lines or higher.

The *Image Time Persistence* is the time taken by the material of the pickup tube's target to return to its original electrical state when you cut off the excitation light. It depends on the nature of the material used, understanding that the greater the amount of light received by the target of the tube, the greater is the persistence. This is the reason why, when a lesser quality tube takes very bright images, you witness a luminous wake on the screen when the camera is moving or when the bright object is moving across the frame. This adverse effect was predicatively accentuated in amateur cameras, so it was not wise to shoot luminous objects or scenes for prolonged periods of time. Several ploys were studied for limiting this shortcoming culminating in the '70s with the introduction of the <u>Anti Comet Tail</u> (*ACT*) device, which was an electronic add-on which significantly reduced (but failed to eliminate) the boring 'comet effect'. This technical expedient was especially employed for the plumbicon-based cameras (and therefore only for the high-end camera market) over the 1980s in a period of fervent innovations in the audiovisual field.

Importantly, a long exposure to blinding light sources brought irreversible damage to the sensitive target of every type of tube with the appearance of fixed and tiny, yet visible, white dots on the screen, or in the worst case a permanent scorch, so shortening the tube's life, that was already relatively limited, being substantially a thermionic valve.

The repair operation to replace a trio of tubes could take several days of work by skilled personnel equipped with dedicated instrumentation at a proportionately high cost. Once the camera was again in operation, the electronic and mechanical calibration of the entire camera had to be re-performed then rechecked frequently.

The *Sensitivity* is the ratio between the electrical signal provided by the camera and the amount of light received through the lens. It can also be expressed as the amount of light needed for which the tube gives a signal of a certain level. It is the equivalent of the sensitivity of photographic emulsions.

Theoretically, when the pickup tube is not excited by any light radiation, it should not generate electric current. In practice, it always produces a current, called *Dark Current* or *Noise Level*. It is the equivalent of the 'veil effect', using photographic emulsion terms. These features, which were a function of the materials used for the construction of the tube, affected the reproduced image quality (especially when it operated at minimum brightness values compatible with the generation of a proper video signal). In summary, disadvantages included the tube's reaction in abrupt changes of illumination, the trailing images in dimly lit scenes, the persistence of the images in bright and isolated points, and the definition of detail. A good camera should have a noise level within limited ranges.

6.2 Solid-state pickup devices

Continuous R&D progress has enabled the creation of a new generation of cameras that no more employ the usual pickup tube to generate video signals but, instead, a solid-state image sensor whose target consists of a large number of tiny photosensitive semiconductor devices perfectly aligned in rows and columns to form a mosaic of picture elements or *Pixels*.

When light strikes these elements, a clocking circuit, synchronized to the television scan rate, drives the image sensor's address decoder that allows each portion of the image to be individually explored, reading the voltage of each pixel, being proportional to the light intensity. The collected voltages so create a video signal without incurring the inaccuracies of a magnetic beam scanning and ensuring an excellent output signal quality.

6.2.1 The CCD (Charge Coupled Device)

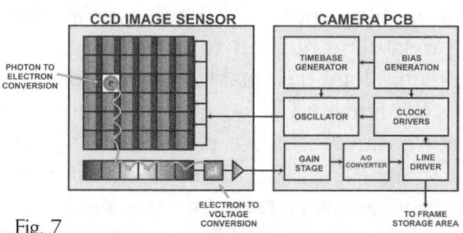

Fig. 7

Fig. 8 - Delta-Doped Charged Coupled Device for ultraviolet and visible detection

The most well known of these sensors has to be the *Charge Coupled Device* (Fig. 7 & 8), shortly *CCD*, based on the concept of the *Bucket Brigade Device* (*BBD*), developed by *Frederik Leonard Johan Sangster* and *Kees Teer* of the Dutch Philips Research Labs in 1969. Cleverly, they soon realized that their invention was also capable of being used as in a solid-state image sensor. Basically, the BBD is a charge-transfer device in which charge packets are transported from one cell (usually a special thin capacitor) to another. In 1970, *William Boyle* and *George Smith* of Bell Labs, in the United States, improved the charge-transfer concept by introducing a transport mechanism from one capacitor to another one.

Nowadays almost all types of digital video or still cameras use the CCD, and even if the earlier devices allowed very low resolutions (the CCD prototype was composed of only 6 pixels!), CCD has now become the most popular technology replacing almost all the other imaging technologies. The working principles of CCD sensors are based on the MOSFET capacitor technology (IDC) ap-

Chapter 6. Video Cameras

plied to an array of very low inductance chip capacitors. Their polycrystalline silicon electrodes have metallic properties but, unlike a conventional MOSFET capacitor, they are quite transparent allowing light to pass straight through and therefore producing electricity from photons incident on the semiconductor.

The metallic electrode is isolated from the MOSFET's P-type semiconductor through a film of silicon dioxide. Applying a positive voltage to the electrode of this structure causes many of the photoelectrons (generated by the decomposition of a photon into an electron-hole) to be concentred under the positive electrode due to electrical attraction. The charge stored in the polycrystalline silicon contact will be proportional to the amount of light hitting the surface of the electrode.

Once stored, the charge representation of incident light must be 'read'. The readout is achieved by changing the voltage applied to the individual terminals of adjacent MOSFET capacitors so as externally transfer each charge between adjacent capacitors. The displacement (or coupling) of charges is obtained through the application of external clock signals to the electrodes of the structure.

Generally clock signals for reading the CCD are provided by a dedicated integrated circuit. Typically the cameras integrate the functions of clock, drivers, control circuits and signal processing into integrated circuits external to the CCD chip. The generation of clock signals for reading is not simple and it requires a specialized chip to provide the correct waveforms and voltage levels. Most cameras operate with external power supplies, namely batteries, and therefore specialised voltage regulators are required to generate the diverse internal voltages required for charge coupling.

Using BBD mechanisms, CCD technology has permitted simple CCD-based delay lines to be created which can replace the old metal/glass delay line needed to correctly decode PAL or SECAM signals in a television set. The same technology has enabled optical-sensitive CCDs with varieties of configurations and uses, such as the linear CCD, so named because the photosensitive cells are placed side by side in a straight line. This CCD is used in scanners, photocopiers, faxes and so-called linear cameras, which are used in some industrial vision systems and to transfer film to video. In order to obtain an image with this type of low cost device a relative movement between the object and the CCD is required.

Fig. 9 - A linear CCD employed into a flatbed scanner

The image must be explored line by line and rebuilt across the movement between the image sensor and the object (Fig. 9).

However for both video and still cameras the most popular CCD image sensors are area sensors meaning they have a rectangular sensitive area with popular differentiation in terms of sizes and total numbers of pixels. A still camera can be equipped with a unique colour CCD presently having 3 to 20 million pixels (or *megapixels*) or even more for most professional still cameras employed, for example in an astronomical field. An amateur video camera needs much lesser resolution. Usually a single 1-megapixel CCD in addition to a special optical filter called the *Bayer Filter* allows capture of only one-third of the colour information on each pixel. The rest of the image information can be obtained through a hardware or software interpolation algorithm.

6.2 Solid-state pickup devices

A semi-professional, a broadcast or a professional portable or studio video camera employs three such 1-megapixel (or even more) CCDs in conjunction with a dichroic mirror and a small spatial offset for each sensor in order to provide a superior image quality. The CCDs for professional video uses are generally sized 1/3 or 2/3 inch with a 4:3 or 16:9 aspect ratio, even if the relative picture definition does not depend so much on the image sensor size but by its total number of pixels.

The cameras that are currently on the market differ according to their cost vs. performance and three popular types of operating transfer modes of the CCD sensor, i.e. *Frame Transfer*, *Interline Transfer* and *Frame Interline Transfer*.

Fig. 10 - Frame Transfer CCD 412 from Fairchild Imagings

The FT (*Frame Transfer*) CCD (Fig. 10) has two separate adjacent sections on the CCD. The first half of the sensor is a grid area sensitive to light and the second half is a grid storage or memory area (masked by a cover of aluminium) used to block any exposure when transferring information relating to 'light reading' from the pixels above them occurs electronically. Such transfer happens during the vertical blanking while the image sensor is illuminated. However a sort of contamination can happen when the charges are moved from the light detector grid to the storage grid since the photosensitive cells are constantly stimulated by the optical image during the short period of time in which the content in the detector is transferred to the memory section. Such contamination will create an artefact called *Vertical Smear* especially evident while framing blinding point source lights and it consists of narrow, unpleasant, unwanted and luminous vertical lights in correspondence smeared along all the screen height.

In order to avoid this, a revolving mechanical shutter synchronized with the vertical blanking interval was mounted in front of the FT CCD so that the light which would otherwise illuminate the sensor during the charge transfer phase is completely blocked. Such an expedient represented a nonsense in the earlier CCD cameras because on one hand it eliminated the smear defect, but on the other hand it introduced an unreliable motorized wheel, i.e. a backwards step towards the old mechanical television! The mechanical shutter was soon replaced by a more efficient

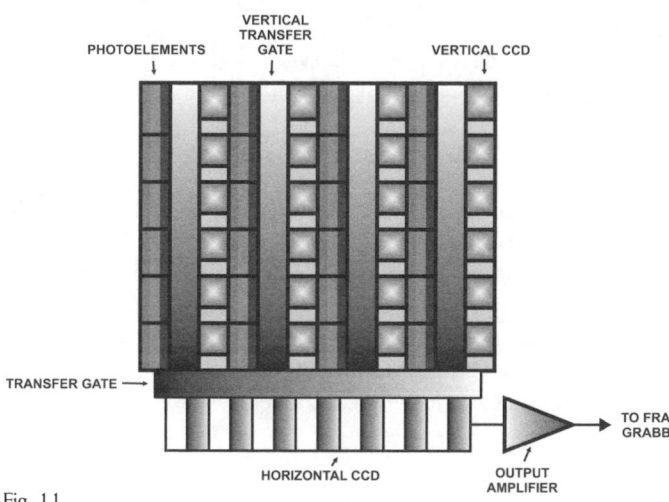

Fig. 11

and modern fully electronic optical shutter based on LCD technology.

For the IT (*Interline Transfer*) CCD (Fig. 11) the approach is completely different from the FT CCD to transfer the charges into the storage area. The storage's elements are protected from light and positioned next to each image item.

During the time period when the image grabbing occurs, the grid of photosensitive cells accumulates charges in proportion to the incident light. Then during the next vertical blanking duration these charges are quickly transferred to the adjacent memory elements protected from light, called the *Interline Storage Registers*. Next, during the subsequent capture, the grid of photosensitive cells, now empty, would be recharged again with the new image while the previous image, read from the memory registers, would be sent to the next stages of the process.

The use of IT CCDs allowed removing the charges holding image information without being contaminated by most unwanted light. However, although attenuated, the problem bound to point source highlights persisted, that is the vertical smear, although it could be almost completely eliminated with the addition of electronic shutters.

The FIT (*Frame Interline Transfer*) CCD combines the advantages of IT CCD with those of the FT CCD and virtually eliminates the problem of vertical smear.

This type of CCD, however, is a very complex device, difficult to built and relatively very expensive. The concept with which the charges are transferred is similar to that of the IT CCD but here there is a double transfer. The charges are first detected by photosensitive elements and subsequently passed into a register memory interline during the vertical blanking and immedi-

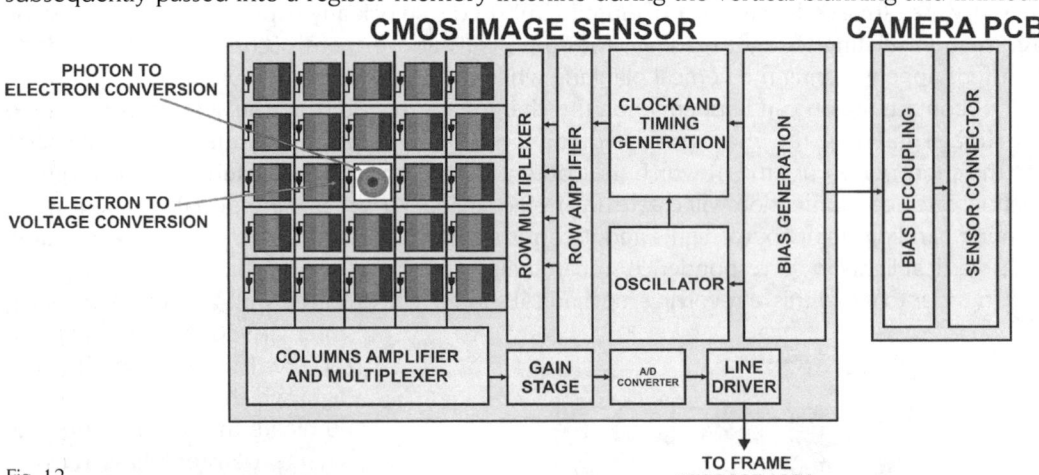

Fig. 12

Fig. 13

ately later transferred to a memory frame grid completely protected from light.

The charges are present in the register interline for a very short time and then the contamination caused by bright light, although present, is greatly diminished. With this CCD the phenomenon of vertical smear is virtually eliminated, i.e. an insignificant level.

6.2.2 CMOS image sensors

Recently, a new image sensor emerged as an alternative to the CCD: the *CMOS* (*Complementary Metal Oxide Semiconductor*) image sensor (Fig. 12 & 13). It provides the integration of additional circuitry on a single chip, lower energy requirements and a more compact design at the expense of image quality and flexibility. CMOS image sensors significantly reduce the cost of the camera because they contain all the logical components required for the camera, being fabricated using the consolidated and popular standard CMOS process with little or zero modification. The architecture of each sensor's element can be of three types: *Passive Pixel Sensor* (*PPS*), *Active Pixel Sensor* (*APS*) and *Digital Pixel Sensor* (*DPS*). A tiny photodiode is the common element in all the three configurations; in the first case (PPS) the signal generated by the photodiode is amplified by only one JFET transistor, in the second case (APS) by usually three JFET transistors (Fig. 14) and in the last case (DPS) the signal is immediately converted into a digital form via an *Analogue-to-Digital Converter* (ADC). The readout is accomplished using a method quite similar to a sequential memory addressing, i.e. for rows and columns.

Each method presents pros and cons. The PPS method, the simplest of all, guarantees a small pixel dimension, easy designing and low cost but suffers from low *Signal-to-Noise Ratio* (SNR), low light sensitivity, non-storage 'destructive' readout and slow readout speed. The APS has larger pixel size, higher SNR and non-destructive and faster readout but results in a more complex design and higher production costs. The DPS has the largest pixel size (due to the use of an ADC for each pixel), lower SNR with respect to APS (but improved via some noise reduction techniques recently introduced) and the possibility to freeze the pixel information before being transferred into a storage media, useful for the larger image size sensors.

CMOS sensors are used for small cameras (e.g. webcams, mobile phone cameras, etc.) even if they are available in larger sizes for low and high-end digital still cameras. Although the applications gap between CCD and CMOS sensors is rapidly decreasing, the image created by the CCD remains significantly better in low lighting. Especially CMOS image devices in the same low-lighting conditions can create images with interferences or distortion but very dark and very noisy.

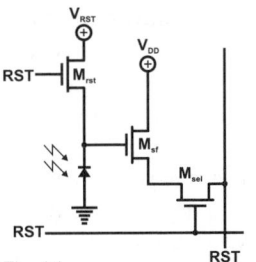
Fig. 14

Just recently Sony (and rumours suggest Panasonic also is working on something similar too) has marketed in its digital still cameras an improved CMOS image sensor, named *Exmor-R*[TM], that would be twice as sensitive to light with respect to standard CMOS image sensors thanks to backlight technology that allows physically shifting the photodiodes above all the electrical connections.

The Italian *Federico Faggin*[2], president and CEO of *Foveon* (just acquired by the Japanese *Sigma Corporation* in November 2008) introduced a family of new revolutionary CMOS image sensors of fourth generation codenamed *X3*. These sensors were inspired by photographic film technology where the sensitive emulsion is composed of three layers which are directly exposed by red, green and blue light. In a Foveon's X3 sensor there are three layers of photodiodes and each one captures one of the three primary colours with the net effect of producing images that are sharper and have significantly reduced image artefacts compared to competing image sensor technologies (Fig. 15). The biggest drawback of X3 sensors

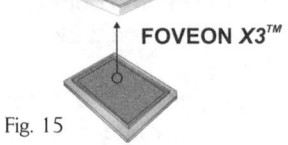
Fig. 15

is the reduced resolution actually available. Although at the time of writing Foveon produces 14-megapixels X3 image sensors, the effective resolution must be divided by three, i.e. by the number of the primary colours. Now that is presently quite inferior to a standard, cheap and diffused 10-megapixel CCD or normal CMOS image sensor embedded in commercial still cameras.

Actually the Foveon X3 sensor is a costly-production device which suffers from some minor (yet acceptable) drawbacks. However, it is based upon a promising image technology which could represent the future of the next CMOS-based image sensors. More information is available on *www.foveon.com*.

6.3 Features common to all Cameras

Fig. 16

There are some characteristics which are common to all cameras, from the tiniest to even broadcast quality. However, depending on the construction or cost, such characteristics could greatly vary from camera to camera.

The *Optical Lens* (referred also as an *Objective Lens*) is a device that forms an image by refraction. It has from one to several single elements of glass (or other transparent materials such as fluorite, quartz or even plastic) having two opposing faces, at least one of which is curved. The comparison with the human eye's lens is evident: although the latter is generally a highly sophisticated optical system it cannot compete with the most complex and professional lens employed, for example, in astronomical (telephoto lens) or medical (optical microscope) fields. Moreover a camera lens can embed important add-on lenses, grouped to form the so-called *Zoom Lens*, which permits you to virtually draw right up to the framed object whilst maintaining it almost in focus. An additional lens can bring the object in focus with precision, being controlled manually. Recently, especially in portable cameras, a device called the *Autofocus* (or automatic focus) can manage this, simply taking it at the centre of the frame.

The zoom lens was invented in 1902 by *Clile C. Allen* but in 1950 *Pierre Angénieux* introduced the *Retrofocus Lens* which, in 1956, permitted the building of compact variable-focus lenses for video uses. Prior to that time the camera operator (better known as the *Cameraman*) had to manually change from one lens to another, often 'on the fly', selecting one of three to six lenses of fixed and variable focal lengths placed on a revolving turret. Nowadays almost all the zoom performance is achieved using a motorized system activated manually by the cameraman via a rocker switch (Fig. 16).

The *Iris* is a variable-aperture diaphragm which adjusts the amount of light entering to expose the camera's target at an appropriate illumination level. A sub-optimal regulation can furnish dark and unsaturated images while an excessive aperture will produce whitish and solarised pictures. Also in order to prevent this some cameras have a motor-driven iris controlled electronically by an *Auto Iris* circuit in addition to a manual adjustment which permits closing the iris completely to minimise accidental damage of the image sensor(s).

If at minimum aperture size the auto iris cannot manage to obtain a proper image due, for example, to an excessively strong illumination, the cameraman must intervene by interposing a

6.3 Features common to all Cameras

Fig. 17

so-called *Neutral Density Filter* (or *ND Filter*) between the lens and the camera's target before the dichroic mirrors. An ND filter is essentially a 'grey' filter, which dims light of all colours, equally. Generally three filters going progressively denser can be collocated on a manually rotated externally accessible wheel to be selected by the operator. You can find the ND filter wheel on professional and semi-professional cameras. Such cameras can take advantage also of some special optical filters such as infrared, anti-aliasing, colour correction and wavelength filters in addition to special effect filters (Fig. 17).

Optical lenses are generally interchangeable between cameras with the same image sensor size of different brands and they follow the so-called *C-Mount* standard, with some exceptions. On amateur and low-cost cameras the optics is fixed and cannot be substituted or, in most cases, replaced if damaged.

A colour can also be expressed in terms of *Colour Temperature,* measured in Kelvin degrees (K). All image sensors are pre-calibrated to give an equal response to all colours at the colour temperature of a tungsten light source, nominally 3,200 K, also called 'warm white colour'. A problem could arise in an outdoor environment with sunlight which gives a dominant blue-violet light, referred as 5,600 K of colour temperature, called 'cold white colour'. Under cloudy weather illumination the colour temperature rises up to 6,500 K due to the exaggerate presence of ultraviolet rays, irrelevant for human vision but detectable by image sensors.

In order to produce correctly-coloured pictures an optical conversion filter can change the spectral range of a light source and has to be interposed between the lens and the camera's image sensors aided also by a special electronic circuit known as *White Balance Adjustment,* set manually by pressing a push-button switch on the camera for some seconds while framing a white object such as a white paper sheet. Nowadays modern cameras are equipped with so-called *Auto White/Black Balance* circuits which facilitate the colour temperature calibration, eliminating the use of a conversion filter. However *Auto White/Black Balance* circuit can be exchanged for manual calibration for some special purposes.

Professional cameras also possess some typical add-ons for improving picture quality among which is the *White Clip Circuit* that prevents the output signals from exceeding a nominal video level of 100 IRE or 1Vpp.

The *Zebra Stripe Level Indicator* generates a 'zebra pattern', visible in the viewfinder only, by which you can judge if picture areas are too highlighted by superimposing the stripes above them.

The *Detail Level* is an image enhancement circuit used in all cameras to improve picture quality by 'overshooting' the edges of the framed objects. An excess of such regulation can suffer from excessive noise on pictures.

The *Gain Switch* is a two or three-position switch which allows you to gain more signal strength when working under very low light conditions, expressed in dB steps. A collateral effect is a progressively noisy picture produced by progressively increasing the gain.

Output Switch allows you to switch either from the normal video picture output or an internal colour-bars generator.

VTR Save/Standby Switch controls the video recorder's start-pause sequence and the turn off and on of the camera.

Audio Level and *Input Controls* permit adjustment, as required, of the audio levels, from external sources (mono or stereo *Line* input) or the on-board microphone (*Mic* input), selectable by a switch.

Time Code Selection inserts a temporal reference code in the vertical blanking, in the form of hours, minutes, seconds and frames (i.e. displayed as '00:00:00:00'), then recorded on media (usually tape) and therefore useful for later making a precise video montage on post-production.

Scene Files are preset data saved in a semiconductor memory within the camera itself. A scene file holds some fine parameters including white balance, white clip and black levels, detail level, camera gain, iris settings and other information. The memorized values can be recalled by pressing the corresponding scene file button on the CCU or the camera itself via a menu visible on the viewfinder. This allows the cameraman instant camera adjustment when moving from one shooting condition to another.

Audio and Video INput/OUTput connectors can diversify from amateur cameras to professional cameras. The former generally uses a proprietary multipin A/V output socket; a special cable splits the relative A/V output signals into separated plugs. Some digital amateur cameras offer a digital video output also (DV out). Professional cameras have separated A/V IN/OUT sockets, generally XLR or RCA jacks for audio signals and BNC or S-Video miniDIN for video signals as well as a bidirectional DV IN/OUT socket.

At this point someone could ask how come that a professional camera would be equipped with an A/V input. The reason is because in 1982, for the first time Sony 'joined' a videotape recorder with a video camera to form the so-called *Camcorder*, a unique standalone piece of portable equipment, based on its *Betacam* video recording system, able to shoot and record footages by employing a single 'independent' operator. In that equipment the recorder section could also be used as a normal storage unit without turning on the whole camera.

Nowadays, all the portable amateur and portable professional cameras (excluding, therefore, studio cameras) are based on the camcorder concept but only professional cameras have the A/V input feature. Prior to the camcorder age, a variety of video camera models were produced for use with separate videotape recorder decks. The *de facto* standard for connecting the two units was then a 10-pin circular connector with a threaded lock ring. Later there was a 14-pin connector developed by Sony that allowed more features on the Sony cameras only. Some manufacturers also built a 10-pin to 14-pin (or vice versa) adapter.

6.4 Some final considerations

The camera is perhaps the most seductive and creative element of the video system. This is undoubtedly at an amateur level, the 'something extra' which makes the difference between the normal viewer and those who, instead, build by themselves 'their television'.

Video electronics is trending to conquer new frontiers towards the molecular world but optical science is maybe reaching the limits of what is possible to diminish the size of a device's pickup lens. One of the most ambitious goals is certainly a cybernetic eye to restore vision to the blind!

Endoscopy relies on small cameras to operate but actually *Capsule Endoscopy* (Fig. 18) probably offers the most progress in video and electronic technology applied to the medical field. It is a relatively small pill which, after having swallowed by a human patient, travels on the food's

6.4 Some final considerations

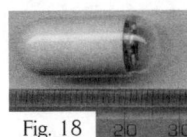

Fig. 18

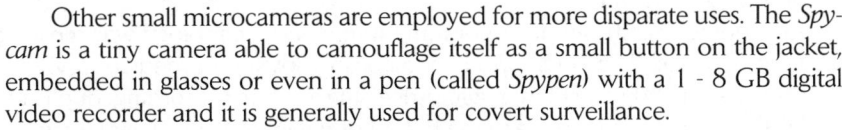

Fig. 19

natural digestive path taking snapshots on the way of the internals of, say, the small intestine. Such photos are sent in real time using a small integrated transmitter and collected by an external wirelessly-linked computer for inspection by specialized medical personnel. A curious note is that this capsule is a throwaway object!

Other small microcameras are employed for more disparate uses. The *Spycam* is a tiny camera able to camouflage itself as a small button on the jacket, embedded in glasses or even in a pen (called *Spypen*) with a 1 - 8 GB digital video recorder and it is generally used for covert surveillance.

The *Webcam* (Fig. 19) is a camera employed on a computer system for surveillance, videoconference, instant messaging, video streaming, WEB TV and point-to-point videophone calls via the internet. The webcam has recently become integrated in laptop and netbook computers.

Also the latest models of mobile phone and PDA (*Personal Digital Assistant*) have been equipped with a small low-quality camera for the purposes of taking snapshots or short footages for personal use or for being published on some video sharing websites.

Doubtless, we can finally affirm that none of the past television discoveries and inventions have become obsolete as, for example, electronic tubes, which are still produced and used in, for example, pickup plumbicon tubes. On the contrary, technologies are constantly changing making it so very difficult to forecast where the future will take video pickup and display devices, which are the scope for which television is born.

We can conclude, having read the last two chapters, that there is a trend to slim screens and miniaturize cameras!

---

[1] Years ago, the late Italian television producer Nanni Loj reported an anecdote where during a television reportage in the Vatican, the then Pope John XXIII appeared in white robes and a technician of the crew immediately 'harangued' him in Roman dialect:
"Holy Father, the white shoots...!".
"What, my son? What do you mean??".
"Shoot, Holy Father, shoot! You cannot wear it on TV... you should change suit...".
So the Pope retired in his rooms and returned with a gray suit. Awed by the technician, the Pope asked:
"And now, my son?".
"Now? You can do what you want, say two prayers, bless some people...".
It was not clear if later that technician was punished or fired.

[2] Federico Faggin is one of the microprocessor's fathers. In 1974 he founded Zilog, the first company dedicated exclusively to the production of microprocessors which introduced in 1976 the Z80 becoming in the eighties the most successful 8-bit microprocessor on market.

7. Video recorder systems

Since early times in the television service, research programs were developed with the aim of achieving a system able to store images from cameras which, once transmitted, would otherwise have been lost.

So another significant accomplishment in video technology was means of storing the large amount of video data on support media also allowing immediate review. Although the film industry was already ahead in image processing, both for recording, duplication and distribution of films, several attempts were made to find or invent an equivalent storage system for television images. The real technological goal was to include immediate review, further separating from the film industry which used chemical-based processes. This was to enable video images to be reviewed, copied or duplicated in large quantity.

Like almost every outstanding technological breakthrough, video recording underwent important and significant changes with the evolution of electronics, from the first tube-based huge prototypes and studio equipment to existing small systems based first on linear and then ultimately on digital devices such as LSI or microprocessor chips.

A reliable storage medium used since the video recording's beginning was (and still is) *Magnetic Tape*, which nowadays is rapidly being supplanted by optical media (DVD) and solid state (flash memories) storage systems, especially for the amateur sector. Currently, in the professional and broadcast sector magnetic tape continues to be used as the preferred video recording medium.

Even if we are speaking about 'video recording', such a term indicates both recording and playback of both images and sounds placed in sync using electronic and mechanical components. This is conceptually similar to a conventional audio tape player/recorder and shares other several similarities too.

However the first video recorders were very oversimplified, i.e. just improved audio recorders 'rearranged' for 'capturing & storing' of a few minutes of low grade and flickering video images only!

Nowadays, however, the most advanced digital video recorders are able to start the recording only when changing motion is detected on successive frames or to record from one to so many frames of video every preset minutes on a standard computer flash memory card, ready to be examined later, even remotely via the internet. This kind of system is especially used in surveillance monitoring and is able to store several days of video sequences. This duration can be hugely increased further by implementing a software compression algorithm!

Quickly let's see now how video recording has evolved, over time, from its beginning to nowadays.

7.1 Video recording chronological history

The first unit able to record an electrically modulated audio signal was credited to *Valdemar Poulson* with the invention of the *Telegraphone* in 1898. It was based on the studies of *Oberlin Smith* who, since 1878, performed experiments on recording such signals on a spooled steel wire, unrolled at constant speed and magnetized by a recording head in accordance with the audio input signal. Later, a magnetic tape was used, i.e. a thin strip of plastic on which was coated an emulsion of steel, ferric or chromium dioxide, or other ferromagnetic compounds (composed of tiny particles of polarisable magnets), stored first as reels and later on cartridges or cassettes. However, the use of such supports had been confined to audio purposes until the early 1950s, due overall to the restricted range of audio frequencies (20-20 kHz of bandwidth) as being unsuitable for video signals, which we know demand higher frequencies (up to 5 MHz).

During the mechanical television age, John Logie Baird proposed, around 1930, his video recorder/player named the *Phonovisor*, based on an aluminium disk medium. It never retailed due to severe and unsolved technical drawbacks. In the same period a home video recording equivalent system called the *Silvatone* was marketed by *Cairns and Morrison Ltd.,* of London. Both disk-based systems recorded and played only a few 30-line images per second, which was probably just adequate for those times. But when the fully electronic television was definitively stabilized first on 405 lines (by BBC) and later on 525 (USA) or 625 lines (Europe) the need for video recording became a problem.

Initially, a compromise solution was the widespread use of photographic film as support for storage and preservation of video footages. The *Film Recorder* was one of the first devices used to record TV images. Its operation was based on the collection of television images from a movie camera focused directly onto a high-definition kinescope screen of a video monitor expressly allocated for this purpose. Such photographic television recording was therefore called *Kinescope Recording*, shortly to be abbreviated by the experts to *Kine*. Yet, because of so many variables of

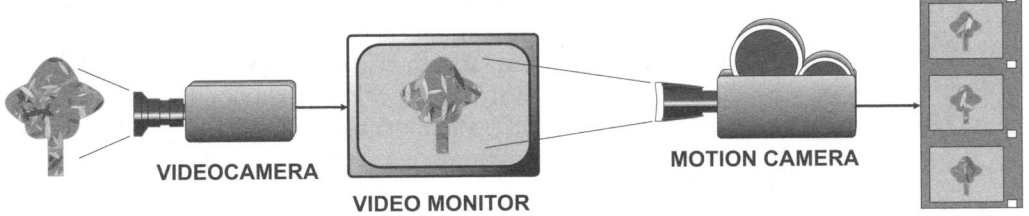

Fig. 1

this electronic-to-film processing, the quality of the result was very poor (Fig. 1).

The film impressed with the movie camera image was next developed in the laboratory. At the end of this process it could be reproduced by a piece of equipment called the *Telecine* which, by means of a video camera framing the projected movie, converted the content of the film back into an electrical video signal, thus ready to be elaborated or transmitted. One can imagine the depth of the signal-to-noise ratio of the final result, which exhibited poor resolution, low-quality greyscale and noisy pictures (Fig. 2)!

Fig. 2

Also this system did not permit the immediate playback feature like a true recording system should do and lacked long recording capabilities due to the limited length of film reels and their high cost. Just a few television events were recorded on this film, and some of these survived and are saved in specialized historical archives.

Since film recording was such a slow and complex procedure, the American companies *Bing Crosby Enterprises* (BCE) and *RCA* focussed their efforts and researchers in early 1950s

Chapter 7. Video recorder systems

Fig. 3

into developing the first usable magnetic tape longitudinal video recorder. This was to be based, in essence, on the principles of magnetic sound recording by using stationary writing and reading heads. It used a long, thin tape about 0.043 mm thick and either ½ inch (about 12mm) wide for colour video or ¼ inch wide for monochrome video, running at a speed of 90 cm/sec. The tape reels were 19 inches diameter with a maximum of fifteen minutes of duration for each one. The colour version utilized a five-track recording, three tracks for primary colours, one for syncs and the last one for the audio signal.

In Europe, also the BBC experimented from 1952 to 1958 on a similar high-speed linear videotape system called the *VERA* (*Vision Electronic Recording Apparatus*), but this and all the other linear systems were soon abandoned in favour of the first real innovation in the field of video recording; the so-called *Transverse Recording System*, codenamed *Quadruplex*. It was employed commercially for the first time in 1956 by Ampex with its *VR-1000* videotape recorder (Fig. 3) and later by RCA with its model *TRT-1A*, the first Quadruplex-compatible video recorder.

Each equipment utilized four small heads mounted on the rim of a disk and placed on an

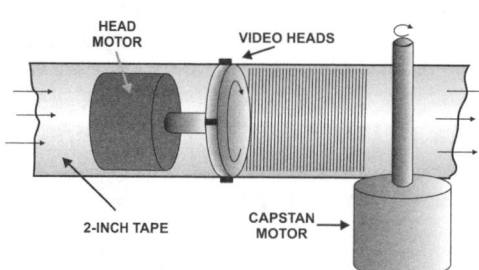

Fig. 4

orthogonal axis between them, i.e. spaced 90 degrees on the disk's border. The head-disk span at constant speed on a plane perpendicular to the tape, the latter being kept in correct position with respect to the head-disk by a sort of 'concave vacuum guide leader'. The recorded tracks, in consequence of the disk rotation and the tape shifting, were placed down parallel with a slight inclination of 0.54 degrees. The tape, which had a width of two inches and a thickness of 0.015 inches, moved at a speed of 38 cm per second, while the heads rotated at 250 revolutions per second (Fig. 4).

Each track recorded by each of the heads contained information of 1/40 of the total television lines that formed a single video frame. Therefore, a complete picture or image was contained in 40 successive tracks, which occupied 15.8 cm of tape. This made it impossible to freeze or slow down images during playback! A very efficient speed control had to guarantee a rock-steady track alignment otherwise, and easily, the video signal could become instable or even disappear.

Despite the technical problems, the transverse recording ensured an excellent image quality, managing to record frequencies up to 6 MHz and a duration of recording up to 1 hour. It had some severe limitations; the tape format was too large and, since the unit was based on thermionic valves, it had a great complexity of circuits and power units (it needed a 3-phase power to operate!) and, therefore it was very heavy (over 500 Kg!) and would have been large.

Undoubtedly, such early video recorders significantly helped the broadcasters because they offered great advantages of magnetic storage over film storage. For example, all the recordings made with Quadruplex-based equipments, being completely a real time process, could be re-

7.1 Video recording chronological history

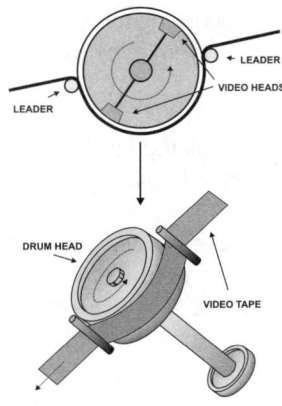

Fig. 5

broadcast either by time-zone, across the US states and in countries outside the US.

It is curious how the 2-inch tape was edited, i.e. very similar to cinema film, using scissors and scotch tape! A special electronic device similar to a cinema moviola detected the invisible gap between two frames, hence, where to cut and join the tape.

The Quadruplex video recorders were adopted only by major television networks because of the high cost of the equipment (45,000 US$, nowadays over 270,000 Euros, but in despite of this high cost about eighty units were sold!) and its maintenance (averagely 200 hours lifetime for video heads!) but they remained in common use up until the early 1980s and the word 'Ampex' became a synonym of 'video recording' among video professionals.

Another important innovation in video recording systems was the so-called *Helical Scan*, so named because the tape is wrapped around rotating recording heads which, while spinning, resemble a section of a spiral. Although there are several variations of this system, generally it employed two heads both for record and playback, placed and fixed diametrically opposed on a cylindrical drum (Fig. 5).

The speed of the drum's rotation, along with the speed and position of the tape with respect to the drum, ensures that the heads record slanted parallel tracks, each containing the information of a half-frame or field. For each rotation of the disc a full image is therefore recorded.

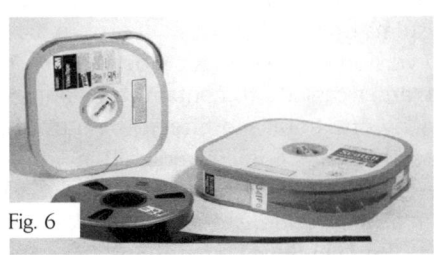

Fig. 6

Since each head records a field, you can stop the tape play, so freezing the image, or you can slow down the playback to play in slow motion. Since the inclination of the tracks was generally of a few degrees with respect to the drum's plane, another advantage was greater exploitation of the tape surface, thus being able to use a narrower tape than in the Quadruplex system.

The head-to-tape relative speed depends not only on the spinning speed of the drum, but also on its diameter. Unfortunately, the various video recording systems that adopted helical scan employed drums of different diameters which made them mutually incompatible, in addition to other things.

Furthermore, each manufacturer chose different drum speeds and inclination (or more precisely the *Azimuth*) of the heads with respect to the tape, size of magnetic tape, and even the technique for loading and winding the tape on the drum, worsening the format's interchangeability.

The helical scan was first introduced by Sony's *EV-200* in 1964 with a transportable monochrome video recorder called *Type 'A'* which used two 8-inch reels which winded a 1-inch tape (Fig. 6). This was

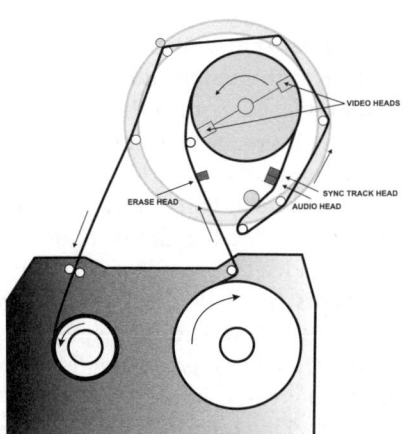

Fig. 7

177

Chapter 7. Video recorder systems

followed successively by the improved version 'B' and 'C'. The time recording limit was always one hour. The 'C' version produced very high quality colour pictures and remained in operation until the end-nineties.

However, another big innovation from Sony was the introduction in 1971 of the first video recorder based on a ¾ inch tape winding between two reels fully encapsulated inside a cartridge (or cassette), codenamed *U-matic* because of the shape of the automatic loading of the tape around the drum, which resembled the form of the 'U' letter (Fig. 7).

Fig. 8

One part of the U-matic development project was planned since 1970 by an agreement between Sony, Matsushita, JVC, Hitachi, Panasonic and five other, non-Japanese, companies, for the unification of the U-matic standard.

So in the following years the U-matic rapidly became a *de facto* standard among professionals and broadcasters, also replacing the movie camera in outdoor uses even if the portable equipment was much heavier. In fact it was, on average, 15 Kg depending on the model and brand and excluding the video camera which was handled in most cases by the cameraman only, while for the recorder another operator was needed. Every cassette, suffixed as 'S' meaning 'Small', had a duration of 20 minutes only, even though *3M* and *BASF* manufactured a 30-minute cassette for portable systems only.

Fig. 9

The studio version of the U-matic video recorder (Fig. 8) employed a larger cassette to contain the tape needed to record and play one hour and fifteen minutes of footage (Fig. 9). In early 1980s Sony introduced an improved version of the U-matic called the *BVU (Broadcast Video U-matic)*, known also as *High-Band*, and partially backward compatible with the *Low-Band* U-matic, and later another improvement called the *BVU Superior Performance (BVU-SP)*.

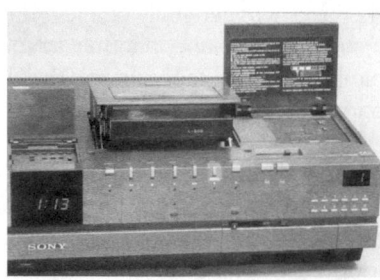
Fig. 10

Nevertheless the great diffusion of video recording was in the amateur field with the introduction in 1975 of Sony's *Betamax* system which led an authentic revolution in the way to conceive the recording means. Just as for the U-matic, the Betamax was so-called for the similarity of the shape of the loading technique of the tape around the head drum which resembled the Greek letter beta (ß). The 'max' suffix gave an idea of 'greatness' (Fig. 10).

Fig.12

Although three years prior, Philips had marketed a proprietary cassette-based video recording system called, simply, the *VCR (Video Cassette Recorder)*, an idiomatic expression which eventually came into common use (Fig. 11), it was never produced in volumes. The Betamax was

178

7.1 Video recording chronological history

the first widely available home video recording system, sold mainly in Japan, the United States and some South American countries. However, just one year later JVC introduced its own video recording system called the *VHS* (*Video Home System*), having some of its parts licensed even by Sony, yet qualitatively inferior to Betamax in everything except the duration of the cassettes (Fig. 12).

Henceforth, an authentic 'video recording format standard war' began, Betamax vs. VHS. JVC licensed the VHS technology to several industries, significantly reducing the equipment's cost compared to Sony which initially backed the Betamax technology, a resounding error! In addition the VHS boom was increased also by the possibility to rent both VHS video recorder and pre-recorded cassettes containing movies most notably from the adult-movie (i.e. pornographic) market.

Soon, VHS recorders spread worldwide and secured the majority of the amateur video market. Sony with Betamax, in contrast, was struggling to forge alliances with other hardware manufacturers. Furthermore, the Betamax's decline was worsened by a lawsuit of Sony vs. *Universal Studios* plus *Walt Disney Company* just few months after the introduction of its video recording system. The movie majors accused Sony, as the first producer of a video recorder, of violating copyright law for audiovisual products. In 1984 Sony won the last grade of trial, but the company image was damaged almost irreparably.

Sony was accused of having introduced *Time Shifting* technology into its video recorder, i.e. the ability to record a TV program on a cassette to be viewed later or to duplicate the program on another tape, a feature that VHS owned from its birth but JVC had never been accused of that!

The battleground of both systems was over several factors: they were rivals in everything! Both employed two very different formats of cassette but inside there was still a ½ inch tape. The Betamax cassette was about one third the size smaller than the VHS and therefore lighter. However, the Betamax contained less tape (and played faster) with consequently less recording time with respect to VHS, the latter allowing longer recordings, most desirable for such as special sport events (Fig 13).

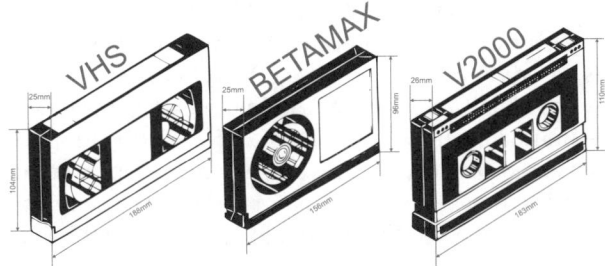

Fig. 13

In the following years for every new idea proposed by Sony for its system, JVC answered back later with the equivalent idea and vice versa; every attempt to improve Betamax for market quota, by now dramatically disquieting, was defeated by the VHS system.

For example, JVC introduced a smaller version of its VHS cassette, suffixed 'C' standing for 'Compact' and fully compatible with the legacy format via a full-size adapter shell (Fig. 14), and smaller portable equipment called, naturally, the *VHS-C Recorder* to counter Sony's *Betamovie*, a camcorder for the consumer market. To retain compatibility with the VHS format, the

Fig. 14

179

Chapter 7. Video recorder systems

rotation speed of VHS-C head drums was increased by 1/3, while the tape wrapping angle was increased to 270 degrees.

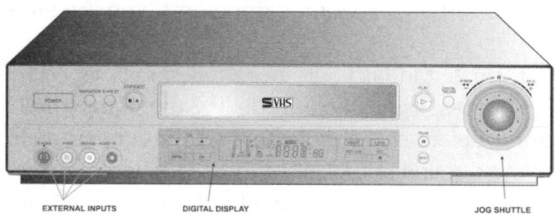

Fig. 15

Furthermore, the introduction of a long duration recording technique by JVC (called the *Long Playing*, *LP*, or the *Extended Playing*, or *EP*), matched a feature available by Betamax too, just like the prompt introduction of hi-fi audio for both systems. Or even a new improved (although unsuccessful) *Super Betamax*, which was promptly replied to by the *Super VHS* (Fig. 15). Sony also produced a high quality audio and video recording system called *Extended Definition Beta* (*ED-Beta*) and later (1982) the professional *Betacam* which became very popular in the broadcast field.

The format war ended in 1988 when Sony began to market its own-designed VHS machines, even though Betamax units continued to be produced until 2002 especially for South America and Japan.

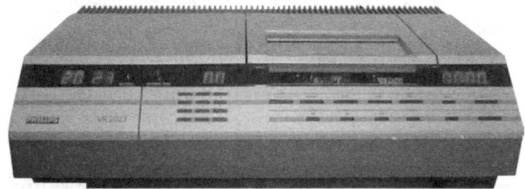

Fig. 16

However, during the Betamax vs. VHS war, a third contender appeared. In February 1980 the Philips Company launched its own video recording standard called the *Video 2000*, shortly *V2000*, which brought several new features, some of which were never matched in consumer Betamax or VHS recorders (Fig. 16). Even though the V2000 was based on a ½ inch tape, the V2000 cassette (called the *VCC*, *Video Compact Cassette*, to follow the Philips-trademarked *Audio Compact Cassette*, and slightly smaller than VHS cassette) had the unusual feature of being 'reversible', i.e. the footages could be recorded on both sides, just like an audio cassette, utilizing one half of the tape's height (i.e. ¼ inch) for each audio-video recording! The time recording limit for a V2000 cassette was four hours per side and it could be even doubled in XL (*eXtra Long*) mode extending the limit to sixteen hours totally (4 hours x 2 sides x 2)!

Rumors circulated in the video community for an incoming compact version of the V2000 cassette to be called the *Video Mini Cassette* (1 hour per side) playable on an existing V2000 equipment by using an adapter, analogous to VHS-C, but it never left the prototype stage.

The V2000 also had a very precise track alignment system, called *Dynamic Track Following* (*DTF*) which, by using a piezoelectric crystal device connected to moving video heads,

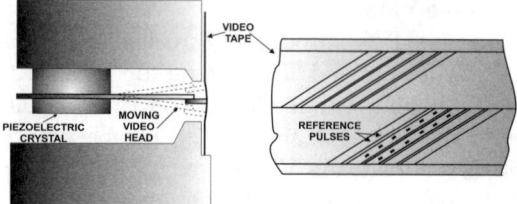

Fig. 17

allowed accurate following of the tracks during the entire playback at any speed. This readily permitted the features of seek and still-image, slow-motion, reverse-playback and fast forward movements without artefacts or 'unstable' video, furnishing an image quality hitherto unknown to the most popular VHS or Betamax units avail-

7.1 Video recording chronological history

Fig. 18

able at those times (Fig. 17). As an additional bonus, DTF allowed automatic rewind and play of the tape with extreme precision at the desired position, using a digital counter. The audio Philips's *DNS* (*Dynamic Noise Suppression*) completed the amazing features of this product.

Grundig also supported the V2000 project but designed for its equipment a different tape loading system and some other mechanical differences which caused some compatibility problems. Being produced practically only by two companies and what's more European (Japan was considered in those times the springboard for new electronic products), the V2000 died in 1988 due overall not only to a tardy market introduction but also to a poor marketing support policy. The result was that confusion and lack of competitiveness led Philips to stop production of V2000 units and a shift to produce only VHS-compatible video recorders since the end of the eighties, instead concentrating its R&D and production efforts into an optical audio support called the *Compact Disc*.

Just as happened with Betamax, not always a smart idea, a big innovation or a superior quality (indubitably in the case of Betamax and V2000 vs. VHS) can easily bend the trend of the market towards one or another technology.

Sony, however, took a partial revenge on VHS, in 1985, by proposing a very successful camcorder based on a new video recording standard, the so-called *Video 8*, where the number '8' meant the height in millimetres of the tape (Fig. 18). The respective cassettes were smaller than VHS-C but with much more recording time (up to 180 minutes versus 45 minutes for VHS-C), as the relative small and light camcorder could even stay in a palm of one hand.

Video 8 was adopted by numerous producers despite its incompatibility with a common domestic VHS deck. Indeed, every Video 8 recording had to be reviewed through the same host camcorder, used as a player, or dumped into a standard VHS cassette thereby losing some image quality.

Fig. 19

Following the Video 8 success, Sony later introduced the *Video Hi8* (short for *High-Band Video8*) to counteract the Super-VHS format, and successively (1999) a digital version called the *Digital 8* (shortly *D8*). Also Hi8 had a discreet success among amateurs and professionals helped by a large availability of equipment and good technical support.

D8, contrarily to expectations, did not encounter the same appreciation from the end users. The available market segment competed with the *MiniDV* digital video recording format (Fig. 19), which had spread worldwide meanwhile, and the latter prevailed ultimately in both the amateur and professional markets. This can seem a paradox because both D8 and MiniDV were based on the *Digital Video* (*DV*) format launched in 1995 by Sony, JVC, Panasonic and other video camera producers. The main reason of the D8 *debacle* was probably due to the employment of a smaller cassette in the MiniDV system than the D8, which permitted smaller design and handier MiniDV camcorders.

Riding on the crest of the digital wave, several professional digital video recording formats

Chapter 7. Video recorder systems

were introduced by different companies especially for the broadcast market. However almost all were mutually incompatible and veered from the Digital Video specifications.

Some examples of these include *Digital VHS* (*D-VHS*), *D9* (known also as *Digital-S*), *DVCAM, DVCPRO, Digital Betacam, Betacam SX, MPEG IMX* and *HDCAM*.

On the other hand, the migration from analogue to digital video format would be inevitable. The link between the two worlds could be considered the *Betacam* system and also the *MII* (so called from the second version of the disastrous 'M' format developed by Panasonic, now obsolete) Interestingly, from a technology viewpoint, the MII employed a so-called <u>C</u>ompressed <u>T</u>ime <u>D</u>ivision <u>M</u>ultiplexing (*CTDM*) technique, an expedient similar to that used in satellite D2-MAC television system, for tape recording the two difference colour components (i.e. R-Y & B-Y) and the luminance at a high quality level on different temporal sequences using a pseudo-digital technique based on the CCD technology (Fig. 20).

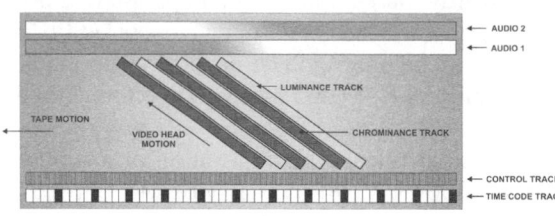

Fig. 20

But the difference between analogue and digital tape recording methods are less than you may think. Helical scan, tape loading and transport mechanisms are quite similar in both recording techniques. The main difference is the way in which the signals are treated and stored in the tape and then 'recalled' from it. We should not forget that every piece of digital equipment must furnish an analogue input and an analogue output as well as digital IN & OUT. The analogue signals will be converted 'on the fly' by internal analogue-to-digital converters prior to be recorded on tape, or vice versa to be replayed.

The optical <u>D</u>igital <u>V</u>ersatile <u>D</u>isk (*DVD*) was a new type of video recording support, born in 1995 to replace tape, which would represent the natural evolution of the optical audio CD. An international agreement among nine electronics producers helped create the DVD specifications upon which all DVD players had to be designed and sold. Initially the DVDs were read-only disks with the purpose overall of selling and renting of both player and movies for this new support, just as originally happened for the VHS format (Fig. 21).

Fig. 21

Later the first DVD Recorders were designed and produced and actually they are now rapidly replacing the video cassette recorders employed in domestic uses being equipped with a digital RF tuner too. The recordable supports were named in diverse ways, according to the brands which introduced and marketed them: *DVD-R* or *DVD+R*. The rewritable medias were called similarly as *DVD-RW* or *DVD+RW*. Depending on the number of optical layers and the level of compression algorithm, each type can record up to eight hours of audio and video.

DVD media is 12 cm in diameter. Smaller (8 cm) recordable (and rewritable) discs have also been produced, called the *MiniDVD-R(W)*, used also in some models of camcorders where the recorded footage could be immediately reviewed via a standard DVD player.

The natural evolution of DVD is the *Blue-ray Disc*, introduced by Sony and another 72 companies in 2004 and winner of a war against the *HD DVD* support, proposed in the same period mainly by Toshiba, Sanyo, NEC and Microsoft. The decision to embed a Blue-ray player in the Sony's *Playstation3* in 2006 was most likely a predominant factor to determine the end of the HD DVD. The *Blue-ray Disc Recorder* at the time of writing is a relatively expensive piece of equipment as well as are the storage discs.

Next innovation in the video recording technology was the so-called <u>H</u>ard <u>D</u>isk <u>D</u>rive Recorder, shortly *HDD Recorder*. In this equipment a standard computer's hard disk is utilized as video and audio storage support driven by suitable software allowing several hours of audio and video to be recorded. Some units are equipped with an interesting feature called the *Timeslip* which permits replaying a scene recorded 30 seconds earlier, therefore slightly delayed in time, while the unit is still in the recording process. This technique also allows the device to record and playback different programs simultaneously.

HDD Recording technology is also employed in some amateur camcorders but the hard drive is generally smaller, fixed and irreplaceable, differently from those of the HDD Recorders. Recently DVD and hard-drive-based units are equipped with digital input and outputs (DV IN/OUT socket known also as *IEEE 1394* or *Firewire*).

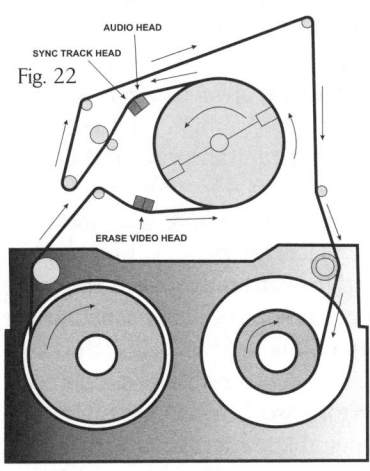

Fig. 22

Currently the video recording trend is towards solid state technology based primarily on computer flash memory media such as *Secure Digital* (*SD*), *Multi Media Card* (*MMC*) or *Compact Flash* (*CF*) and employed successfully in a new generation of fully solid-state digital video camcorders. Depending on the single chip capacity and the efficiency of the compression algorithm, each support is able to record hundreds of minutes of footage, even at *HD* (<u>H</u>igh <u>D</u>efinition) quality.

These digital equipments are rapidly conquering the amateur market to the detriment of all the previous analogue storage technologies. As a consequence, the production of VHS cassettes and relative stand-alone recorders was stopped in 2008. However, JVC and other manufactures still continue to produce combinations made up of a VHS recorder or a DVD player in conjunction with a TV set, and a VHS Recorder plus a DVD player in one unique deck. In 2009 Panasonic has begun to market no less than a dual VHS-Blue ray player!

7.2 Video cassette recorder mechanisms

If on one hand the helical scan solved many problems about video tape recording, on the other hand other drawbacks were introduced.

The first complication was mechanical. The tape threading was quite difficult to implement overall due to the fact that the tape is totally enclosed in a cassette, inaccessible by hand and also protected from dust and light. Several variants of tape threading were studied, based essentially on 'U' and 'ß' loading techniques (Fig. 22), used for U-matic and Betamax respectively, which pulled out quite a large quantity of tape from the cassette. The VHS system implemented a smart loading technique known as 'M' because it resembled the 'M' letter, which pulls out less tape from the cassette (Fig. 23).

Chapter 7. Video recorder systems

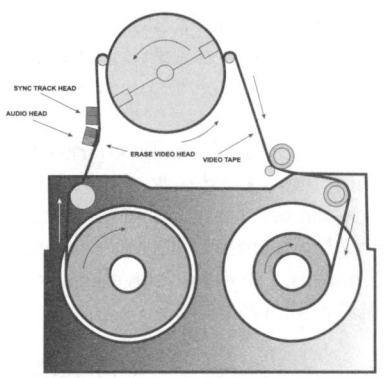

Fig. 23

The task of the tape threading is to open the cover door of the cassette, delicately extract the tape from within, then wind it up around the video head drum at precise locations and height, travelling across capstan, pinch-roller, audio heads and some guide leaders, finally transporting the tape at a constant and exact speed. When the playback, or recording is terminated, the eject of the cassette has to be performed by reinserting the tape inside the cassette using the same system from its operating position in a direction opposite the first and using the same delicacy, without damaging or breaking the tape.

Therefore the tape handling system interacts with the electronics of the video recorder in order to perform the usual set of functions available on VCR units which include fast forward and reverse, play, record, pause/still, slow playing, frame advancing and stop. In addition there are some differences between the threading techniques of the various recording systems. For example in the 'U' and 'M' loading systems, the tape's fast reverse and forward movements are performed by rewinding the tape completely onto the cassette in order to protect the fragile video heads and for accelerating the searching operation. In the Betamax units the tape always travelled across the transport mechanism, included head drum, although slightly released at high speed seeking.

Recent developments have adopted this last solution for all the tape transport mechanisms included digital. This ensures recording of precise reference points or a *Timecode* signal along all the tape's length, useful in the editing, insert, assembly, cut, jump, search, etc., operations on the footages. In addition an optical sensor, placed in a strategic place, guarantees the presence of the cassette in play, record and stop mode. In fast reverse and forward movements the same sensor also detects the end or the beginning of the tape, via transparent leaders attached at the tape's extremities, by immediately stopping the operation otherwise the tape as well as the same transport mechanism could be damaged seriously. Curiously, early recorders utilized a detector comprising a tiny tungsten lamp and a small photocell to perform such a control operation. When the lamp burnt out, the unit become unusable (yet repairable). Successively, the lamp was upgraded to a more reliable infrared LED.

At this point, we can therefore affirm that different and complex movements are carried out by the tape handling system in order to achieve the functions mentioned above. Now you can imagine how much research and development effort has been perpetrated to implement such technology properly!

The old technology is still used in professional and broadcast field (e.g. in Betacam camcorders and relative video montage systems) as well as in the amateur field (VHS, MiniDV, etc.) and nowadays is often still fully working.

Before the tape threading action, there is another delicate operation to perform. The cassette obviously had to be inserted on an aperture of the recorder sized exactly to the cassette's dimension and with some expedients to avoid accidental wrong insertions. These last expedients are absent in the V2000 because its cassette could be inserted on both sides, as previously mentioned. A proper write protection switch detects if the cassette is ready to be played/recorded or played only.

The earlier video recorders had a manual cassette insertion, via a sort of 'ejection cassette

7.2 Video cassette recorder mechanisms

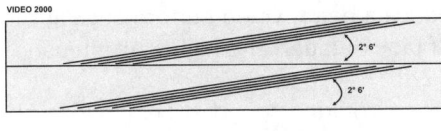

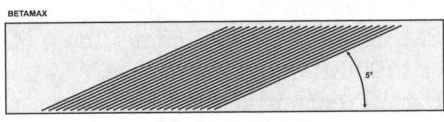

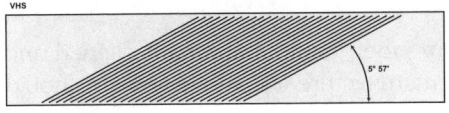

Fig. 24

holder' handled by a specific lever. Later, all recorders were equipped with an automatic insert/eject servo system included in the cassette holder and electronically driven together with all other transport operations via push buttons. Some sensors ensured the perfect and successful insertion of the cassette inside the recorder. Any problems detected during any operation would cause the immediate stop and if possible automatic eject of the cassette.

No less important is the video spinning head drum system which is responsible for the optimum signal translation to/from the recorded material. This critical and delicate part of the video recorder is accurately driven by very sophisticated control circuits, so precise that a variation of just a few thousandths of a degree in the azimuth height (Fig. 24) or speed tape instability could cause serious troubles to the video recovery!

In order to achieve this, another track is recorded along the tape in addition to the inclined video tracks with information stating the exact positions of video tracks. This is nothing more than a 25 Hz square wave (29.97 Hz for NTSC) written during the recording, but it is then used to determine when to stop the readout of video information from one head and switch to read from the second head while they are spinning. In addition, this signal controls the speed of the cylinder head drum motor (see later).

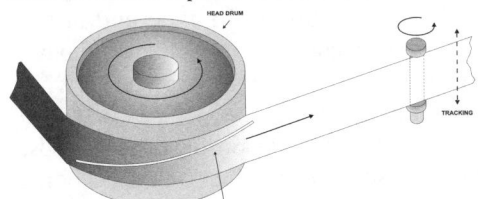

Fig. 25

Fig. 26

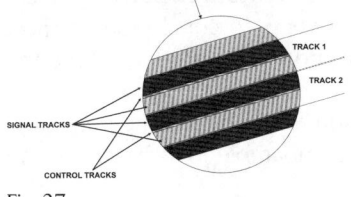

Fig. 27

Furthermore this 'sync track' feeds an *Automatic Tracking Circuit* which ensures perfect alignment of the video heads to the video tracks, by generating a track-locked feedback control signal even if the tapes are recorded by different units (Fig. 25). A weak sync track signal (due to demagnetized or damaged tape, or other reasons) causes loss of video sync stability, 'jumping' frames and other drawbacks. A manual tracking jog could solve the problem in most cases (Fig. 26). The V2000 (other systems followed later) interposed every two video tracks with another parallel 'control' track for improving the head-to-track alignment by dynamically controlling micro movements of the heads (dynamic tracking) so to get a perfectly locked-video in reproduction (Fig. 27).

Since the heads must spin skewed with respect to the tape and are situated on a cylindrical drum, suitable values had to be chosen for the head's inclination degrees and the drum diameters of the various recording systems. Depending on these values, in addition to the tape transport speed, the video quality changes accordingly. A faster transport speed, a lower azimuth of the video tracks and a larger drum dia-

Chapter 7. Video recorder systems

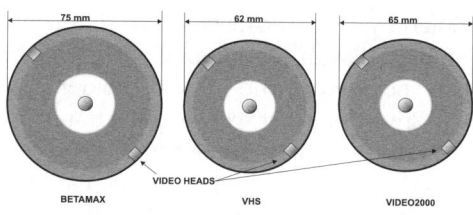

Fig. 28

meter produce a better video performance at the expense of tape duration and higher maintenance costs (Fig. 28).

The drum rotating speed is kept rigorously fixed and depends on the television standard: 1798.2 RPM for NTSC (29.97 frames/sec x 60 seconds) or 1500 RPM for PAL and SECAM (25 frames/sec x 60 seconds). Digital recording systems have different speeds depending on their specifics.

The video heads are considered the heart of every video cassette recorder, aligned and locked into place on the cylinder drum at the point of manufacture and this alignment should never be tampered with.

These microscopic transducers are generally made of ferrite, an extremely hard and also very fragile ceramic material, and they are generally mounted at the lower side of the rotating cylindrical drum. The heads are shaped like a 'C', with a microscopic gap approximately 1 micron wide. The head gap is filled with non-magnetic material in order to push the magnetic field outside the head and inside the tape and also to prevent the gap from collecting debris. Some windings of thin copper wire form the coil of an electromagnet which completes the video head (Fig. 29).

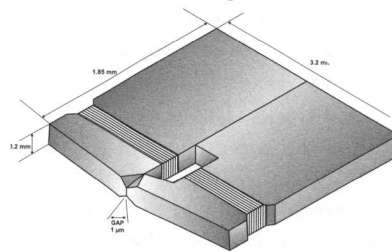

Fig. 29

Two heads are sufficient to perform any function of reproduction, recording or search. Usually the heads are mounted spaced exactly 180 degrees opposite of each other on the cylindrical drum. In the 4-head (or 3 or 5-head) recorders various combinations of heads are used for each function in order to optimize the quality of video recording or playback. In 2-head recorders a pair of heads optimized in width and other characteristics are used. However, there the heads are not mounted exactly at 180 degrees; this arrangement offers a perfect switching signal between them, helped by a video delay line to properly align the two video fields in a video frame but also complicates controlling of the heads.

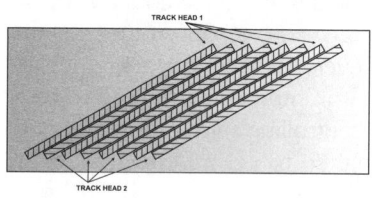

Fig. 30

In order to diminish contamination between adjacent video tracks, the head gaps are differently and oppositely inclined by some degrees, a feature implemented by twisting the two arms of the 'C' during the construction process (Fig. 30).

Another pair of opposing heads is required for hi-fi audio, and a further head is present if the unit has a 'flying' erase head, usually being of double width and able to delete a couple of tracks (which compose two fields or frame) in one sweep. In conclusion, in most recent analogue video recorders there can be up to 7 (or even more) heads which share the space on the cylindrical drum along its entire circumference!

Inside an analogue video recorder there is also a linear fixed audio head. The very slow speed of the tape means that the audio quality is not minimally comparable to a cheap cassette audio player even at standard playback.

186

When hi-fi audio was introduced and recorded to tape by using a separate set of flying heads (rather like the video heads), the 'low quality' audio head remained for backward compatibility with older recordings. Curiously, the Video 8 system and its successive improvements were designed without any linear audio head.

Hi-Fi audio quality was excellent having a frequency response, signal to noise ratio and dynamic range quasi-comparable to a Compact Disc. Some radio stations, in fact, using a 4-hour video cassette in LP or EP modes were able to simply record up to 12 hours of stereo music at extremely low cost, ignoring the video signal at its input even if some recorders would require a suitable video signal just to stabilise the tape speed.

7.3 Magnetic playback and recording techniques

Reliable recording technology based on magnetic tape is still relatively new, with the first tape machines appearing just before the Second World War. Tape-based data recorders are also used on aircraft (the so-called *Black Box*). Although it is relatively easy to magnetize a magnetic tape, the accurate recording of something intelligible and acceptable is not so easy, mainly due to a property of the magnetic materials called *Hysteresis* (Fig. 31). In order to obtain a valid registration pattern, magnetic tape has to be kept at an ambient level of magnetization. The changing of such magnetization on the tape via a modulated signal represents the intelligibility of the recording. However, the hysteresis curve always produces a distortion of the magnetising signal.

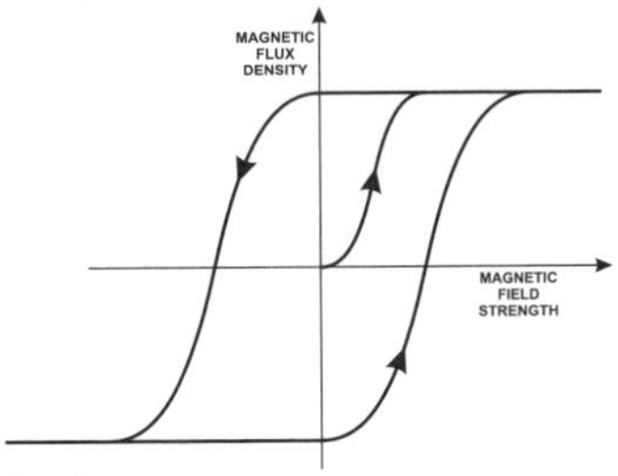

The solution to this problem was the application of an elevated high-frequency carrier signal, amplitude modulated by the audio signal to finally feed the recording head. This carrier is called the *Bias* and contains the energy necessary to overcome the hysteresis. The audio modulation causes the residual magnetism to vary according to the original signal. After the hysteresis has been exceeded, the voltage level of the packets needed to impress a signal on the tape is still quite large. It is not unusual to apply a 100 Vpp at a recording head to store only a few hundred of millivolts on the tape!

Fig. 31

If on one hand the problems of recording an audio signal were solved brilliantly, say by using some further resourcefulness, such as the discovery of new magnetic emulsions for the tape and better manufacturing of the heads, video recording was still to be a challenge.

The first of the problems to be solved was the low frequency response range then obtained from an audio tape. For an audio signal the bandwidth is 20Hz-20kHz, and non-linearities of the magnetic tape could be easily compensated with relatively simple electronics. But for a 5MHz video bandwidth the same approach yielded inadequate results, producing low detail

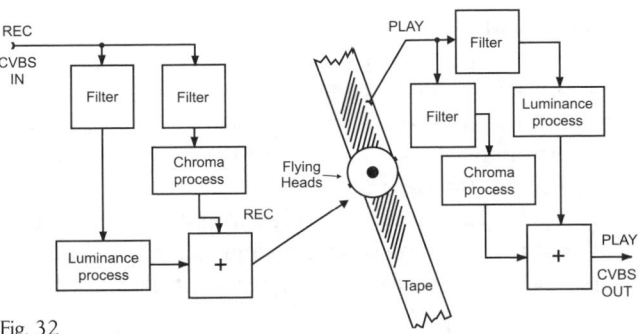

Fig. 32

and noisy pictures, and no equalizer ever invented could readily correct this gain imbalance! Even if transverse recording, helical scan, tinier heads and better tape quality have improved video tape recording globally, a definitive solution was the employment of a carrier modulated by the video signal in frequency rather than amplitude to feed the flying heads. The carrier frequency was chosen to be near the upper limit of the tape's response, while the modulating signal had even to be pre-equalized to improve the linearity (Fig. 32).

Such resourcefulness brought some advantages. Firstly, a reduction in the high frequency noise in the picture was apparent. Secondly, the bias could be eliminated since the FM modulated carrier is always of constant-amplitude, thus the signal could be recorded at the saturation level of the tape.

Before FM modulating the heads signal via a carrier, the video signal must undergo a separation process into its two main components, i.e. luminance and chrominance, which follow two different paths. Luminance has to be separated from chrominance via a low-pass filter limiting top-frequencies to approximately 3MHz, depending on the standard. Chrominance travels through a bandpass filter, then down-converted via a heterodyne method and a local oscillator to be limited to a 0.5-MHz-wide double sideband signal which will feed the carrier directly. The nominal frequency of the carrier depends on the recording standard specifications and television system. Generally it is around 650 kHz.

Finally both processed components are added to the carrier now modulated in frequency by luminance and chrominance ready for recording onto tape (Fig. 33). Video recording stan-

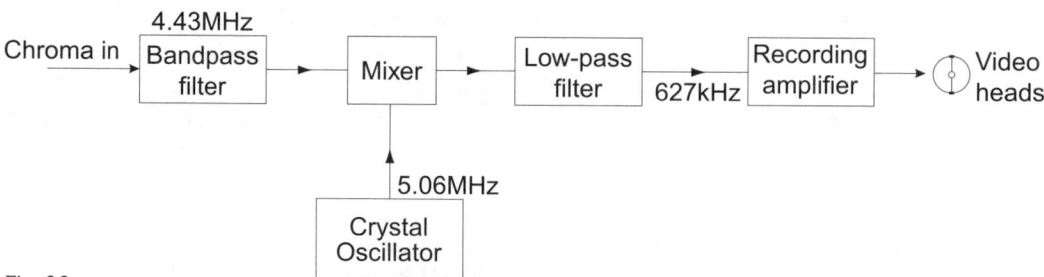

Fig. 33

dard systems which provide separated Y/C components from default, such as S-VHS or Hi8, clear all the separation processes to drive the FM carrier and so producing an improved image quality. Obviously in playback mode the generation of the initial video components follows a reverse demodulation process.

The FM drift keeps track of the luminance variations. The lowest frequency value generally represents the sync level and the highest the white peak level. In the middle is recorded all the greyscale video information, included the blanking level. Since the maximum FM deviation is limited within a specific range, depending on the standard recording specifications, it is clear that the different frequencies chosen for sync tip, black and white peak levels can deliver very different quality of video signals between different standards.

Furthermore, the video signal obtained by magnetic playback suffers from the relative instability of the mechanical tape transport, so that the sync timebase is neither stable nor accurate, since the sync pulses are recorded on tape and not generated by a crystal-controlled oscillator, as happens inside a camera.

In amateur use, where the video recorder is connected to a monitor or a television receiver, the inaccurate timebase does not virtually create any problems of vision. In broadcasting field, where the syncs must be highly stable and clean, the video recorder is equipped with an apparatus called the *Time Base Corrector* which tries to correct all timebase drift. Some TBCs are equipped with a circuit called the *DOC* (*Dropout Compensator*) which eliminates the tape's imperfections (just called *Dropouts*) replacing the damaged scan line with a portion of the line scanned just before it.

7.4 Footage transferring

The transfer of footages from a video recorder to another one is simple, just as it is possible to do with audio transfer, yet without the same accuracy. Some cables with respective sockets connected from source to destination, obviously audio/video outputs to audio/video inputs, are just the minimum for obtaining an 'acceptable' result. Here we are using the term 'acceptable' and not 'good' or 'excellent' for good reasons.

Firstly, the composite video, as we have seen in previous chapters, is a complex signal having fixed timebase pulses and frequencies (syncs, bursts, colour subcarrier, etc.) and has to be transferred point by point unscathed, as much as possible, from tape to tape and this is virtually impossible to achieve. Secondly, such composite video undergoes another modulating-demodulating signal process both in the player and the recorder devices which will further degrade the video signal. Thirdly, depending on the specific video recording standard, the quality of copied material can vary greatly, even among equipment or models of the same brand belonging to the same standards due to the innumerable mechanical and electronic differences between themselves!

These are the major drawbacks of analogue video with respect to digital. Each time that an analogue video signal transits across different equipment, it suffers the consequences of their internal processing stages, introducing noise, phase distortion, high-frequency cut-offs and other curtailments. In a certain sense, a fully digital video transfer between digital units always gives better results than an analogue transfers.

In addition, the boom of pre-recorded videocassettes also increased the quantity of low-quality movies rented out to people. Many an unscrupulous renter, to avoid any possible damage to the renting movie, hardly consigned to the customer the master copy (original), but a copy of it. Furthermore the same customer could take (illegally) another copy of the movie for reviewing it later or other purposes. Since most people possessed amateur VCRs, a 1st or 2nd (or even more) generation copy was visibly and clearly of worse quality. The result was that the video presented poorly-defined and noisy pictures, sync instability and even the audio desynchronized from video!

Analogue video duplication was also problematic in the professional sector. However, professional systems, as the U-matic, or later the S-VHS, additionally employed a technique that used a dedicated cable carrying only the down-converted chrominance to bypass several chrominance stages which improved the quality of copying. Betacam video montage systems gave (and still give...) the best video performances in duplicating footages using YUV component signals or even through the so-called *SDI* (*Serial Digital Interface*) with embedded audio, which allowed a virtually infinite number of generations, with no loss of quality.

The entertainment industry globally which duplicated volumes of audio and video works

Chapter 7. Video recorder systems

ensured that the quality of the recorded product was very high by employing stringent quality control procedures and expensive duplication rooms each containing several parallel-driven professional ½ inch tape recorders.

In the amateur field there was (and still is...) not much more available to do to limit the picture degradation while copying.

In order to counteract copy degradation some 'Picture Enhancer' equipment appeared on the market. Even though such circuits do not perform miracles, by using an enhancer some quality improvements may appear in the copy of 1st generation video footage.

One of these circuits is visible in Fig. 34 which via two RCL filters, one being centred on 1.2MHz and the other one to 2.1MHz, performs sharpness and detail picture enhancement of variable strength by means of user adjustable potentiometers. In addition, a fader stage allows

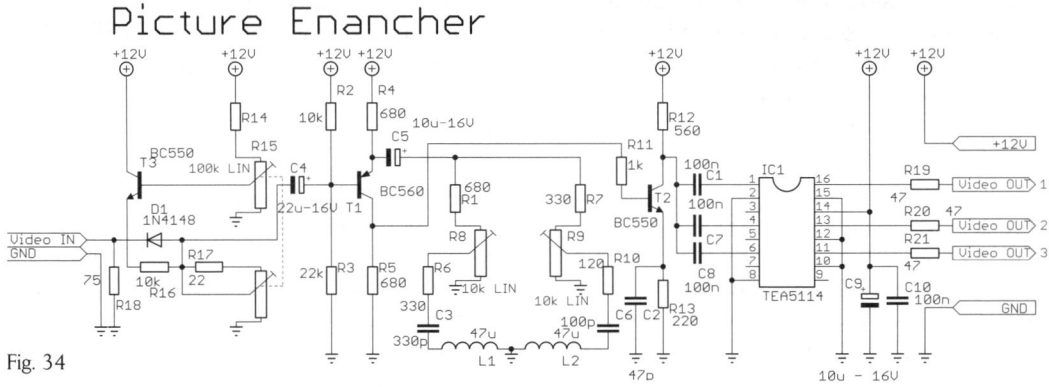

Fig. 34

manual controlled fade-to-black of the video signal using a dual potentiometer. Moreover, the video signal is buffered to drive up to three outputs such as video recorders or TV monitors.

The picture enhancer is mainly based on just three transistors and an IC. The first stage composed by T3 (BC550 or equivalent device), R15, D1, C4 and some resistors, performs the video fade-to-black operation. R15 has the functions of diminishing the video signal level and contemporarily driving the base of the transistor T3, used as an electronically variable resistor. D1 ensures that when both wipers of R15 are grounded, the synchronisms pass through, being of the same level as the diode's junction voltage (namely 0.5V). T1 (BC560), R8, R9 and two RCL filters composed by R6-C3-L1 and R10-C6-L2 form the picture enhancer. In particular, the former RCL filter is the sharpness feature, adjustable via R8, while the latter RCL filter is the detail feature, adjustable via R9. Calibration is empirical, the optimum settings are judged by viewing the recorded material.

T2 (BC550) is a pre-buffer/inverter, which, via C1-C7-C8 drives the IC1 (TEA5114 from ST Microelectronics), a RGB switching integrated circuit capable of driving a 150-ohmload on each output, here used as a triple buffer with a common input. Moreover, IC1's input signal black level is tied to the same reference voltage on each input in order to have no differential voltage between outputs. An AC output signal higher than 2Vpp makes gain decrease slowly down to 0 dB to protect the TV-set's video amplifier from saturation. Such a feature is not required in recent video recorders, these being equipped with a circuit which automatically manages to compensate every difference in the voltage input and stabilizing it at nominally 1Vpp. Such a circuit is called the _Automatic Gain Control_ (ACG) upon which anticopy protections are based.

(see paragraph ¤5 for more information).

The picture enhancer circuit needs a single +12V power supply and it is about 200 mA at full load. The bandwidth is over 20 MHz and the outputs are short-circuit protected.

7.5 Analogue Anticopy Protection

Just like software houses protect their intellectual property (programs, operating systems, etc.) using methods more or less refined (serial numbers, activation by phone, dongles, etc.), film producers, protecting millions of dollars of investment in a movie, introduced in VHS and DVD a special _Analogue Protection System_ (*APS*), sometimes also called the *Copyguard*. In principle, film producers have the right to obtain a fair return on their investment, but the copy protection is designed only to guard against casual copying. The reality of that goal was thus to 'keep the honest people honest'.

This copy-protection system was developed for the first time by *Macrovision*™ in the '90s, practically when thousands of movie titles on videotape were produced for the growing market of home cinema.

Also computer video cards with composite or S-Video (Y/C) output must use APS when these cards play a DVD movie. Inexpensive computer video capture cards have generally the same philosophy: when a Macrovision™ signal is detected, the footage capturing is stopped immediately and a copyright message is displayed instead.

For the NTSC market, Macrovision™ added a *Level 2* of anticopy pulses called the *Colorstripe* pulses which cause variations in the colour subcarrier of a composite analogue video signal, designed to disturb the subcarrier circuit of analogue VCRs. This technique creates a rapidly modulated colour burst signal, with the optional addition of 2-line or 4-line Colorstripe pulses (Fig. 35). On PAL systems this technique is not achievable because it affects the phase of the colour burst. As we know, a PAL colour burst switches its phase line after line, otherwise no PAL user will ever see any movie clearly!

A curious thing is how Macrovision™ is implemented on DVD movies. The disks themselves contain 'trigger bits' telling the Macrovision™ certified MPEG decoder chip of a DVD player whether or not to enable Macrovision™ APS and at what level: *AGC* (so called because it influences the VCR's *Automatic Gain Control*) -> trigger bits 1, AGC + 2-stripe -> trigger bits 2, AGC + 4-stripe -> trigger bits 3.

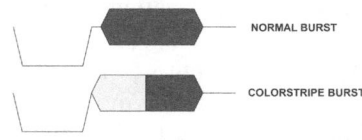

Fig. 35

Satellite and TV-Cable Networks sometimes transmit 'trigger-bitted' footages, incorporated into the stream delivered to a consumer. The *Set Top Box* decoder actives the relative protection level accordingly to the trigger bits detected. For PAL systems, AGC only trigger bit 1 can be enabled.

The triggers occur about twice a second, which allows fine control over what part of the video has to be protected. The DVD producer decides what amount of copy protection to enable and then pays Macrovision™ royalties accordingly. Some DVD disks are Macrovision™ protected and some are not but some cheap DVD players do not activate APS in any case!

The situation becomes further complicated when the system used to prevent copying involves the normal video signal by adding additional pulses, some of them variable-sized and 'dancing'. However, it often also stops many TV sets and video projectors from displaying stable

Video recorder systems

images during the legitimate vision.

In particular, these extra pulses can cause problems with large-screen TV sets which, to reduce flicker, display images at 100 frames per second (100Hz) and also with video projectors which, for improving the image quality, perform a doubling of lines and pixels. They may also cause problems with older television sets.

The only way to achieve steady images on these TV sets is to remove these pulses somehow. The idea is 'to clean up' the video and let the synchronization circuits do the normal work without interference. And this is exactly what we have a good mind to do.

7.5.1 How does the anticopy system work?

Before you take a look at the circuit diagram, it can help to understand how it works by explaining what the anticopy system we are trying to remove does.

The Macrovision™ system adds three main groups of pulses into the video signal, two of them being combined together. First, there are the 'dancing' pulses, immediately after the vertical sync pulse inside 14 (normally) black lines during the vertical blanking interval. In each of these lines, immediately after the colour burst, seven extra pseudo-horizontal sync pulses are added, each of which is immediately followed by short bars whose amplitude cycles slowly (or 'dances') from white-peak level (and even over) for 23 seconds to blanking level for 5 seconds, usually in two or three groups.

In theory, these pulses should not disturb the functioning of the sync separator circuit in a TV set or video projector, but are intended to cause chaos in the AGC circuitry of a video recorder or should block the recording in the DVD Recorder, DV cameras and almost any other digital recorder. In particular, the extra sync pulses should mess up video synchronization, while the bars mislead the various AGC video circuit causing dark/light cycling, stripes of colour, distortion, rolling and black & white pictures. All this actually works, but the devastation is not just limited to video recorders!

The remaining series of pulses are called *EOF* (*End Of Field*) pulses. These consist of a group of narrow positive pulses added into about six lines located at the very low bottom of the image and replacing the colour burst, i.e. they are inserted immediately after the horizontal sync pulses of those six lines. Indeed, these pulses 'push up' the colour burst of those lines over the white level, so that the black level and colour synchronization circuits of a VCR are chaffed

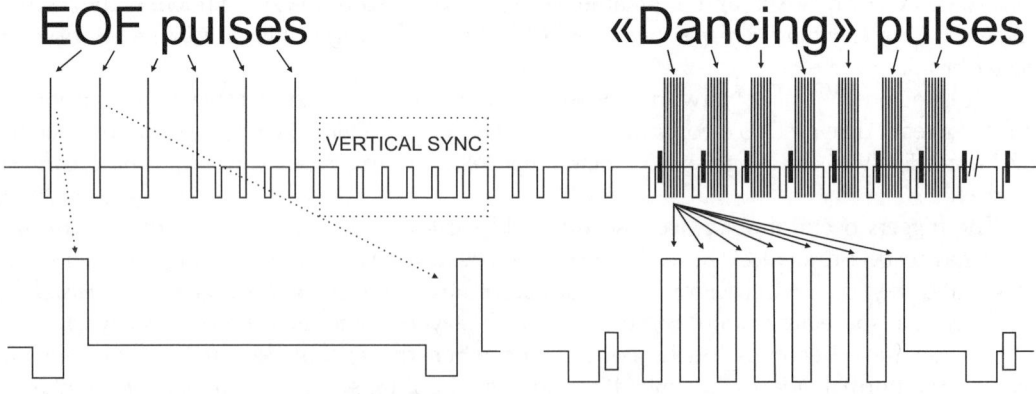

Fig. 36

7.5 Analogue Anticopy Protection

again (Fig. 36).

As previously outlined, Colorstripe pulses are not present on PAL video signal and so there is no significance in trying to remove them.

7.5.2 How to eliminate protection pulses

The internet is crowded with many free diagrams or circuits and commercial products for removing the anticopy signals, variously described in modes, but all share the same problem: they are too complicated or/and expensive to purchase. They range from a simple circuit that uses a programmed EEPROM or a PIC microcontroller, or that uses about a dozen integrated circuits and several discrete passive components. Some do not eliminate all the anticopy signals (i.e. just removing the main ones) and some degrade the video signal when it passes through. Worse, some do not work at all!

As a result of three years of study, especially in 2003, I named this circuit **MK2003 Video Stabilizer** and it boasts of being the smallest PAL-only Macrovision™ eliminator in the world! It is fairly straightforward and is based on six low-cost readily available ICs, and about twenty-five passive components. It also has the feature of being able to bypass the video signal when no protection signal is present (and therefore making no sense to process it), and also offers S-Video input and output. It has a power plug socket to connect a common external power supply also not required to be voltage stabilized. Wherever practical I chose ICs packaged in SMD technology and doing so, the ultimate size of the printed circuit board has been reduced to a bare minimum: in fact its dimensions are only 35x65mm!

Note that the S-Video chrominance signal is not actually processed by the 'filtering' circuitry of the stabilizer (it does not need this), and it must be passed through: in fact pin 4 of both IN and OUT mini-DIN connectors with S-Video signals are interconnected directly. Also note that both composite video input and output respectively, are related to luminance input and output of the mini-DIN sockets (pin 3). In other words this circuit will not convert from composite video to S-Video and vice versa. In summary, use only with one type of signal at a time, without mixing them.

7.5.3 Description of the circuit

The essential video sync separator IC1 (LM1881N) extracts from the composite video all the clock pulses needed to generate the necessary control signals needed to eliminate the Macrovision™ pulses. R2 and C2 from pin 6 of IC1 to ground set the chip's internal timing circuitry for the most accurate and stable sync separation (Fig. 37).

However from IC1 we only need three signals. From pin 1, we get a negative-going composite sync signal (CS), while from pin 3 we get similarly negative-going vertical sync pulses of about 230µs wide (VS). Finally, from pin 5, we get negative-going narrow pulses (about 2.5 µs) which correspond with the colour BURST.

These three signals are then processed by the four non-retriggerable precision monostables in IC4 and IC5, both CD4538D, and by IC3 (CD4001D), a quad 2-input NOR gate.

IC4 cancels the main Macrovision™ pulses, those troublesome 'dancing' ones, while IC5 eliminates EOF pulses. In particular, monostable IC4A is used to produce at its output Q (pin 6) a positive pulse about 1.2 ms long (based on the timing obtained from R4 & C3), starting at the end of the vertical sync pulse from IC1/pin 3, i.e. from its rising edge. During this pulse IC4B will be allowed to play its role because previously its clear pin (pin 13) was lowered to

Video recorder systems

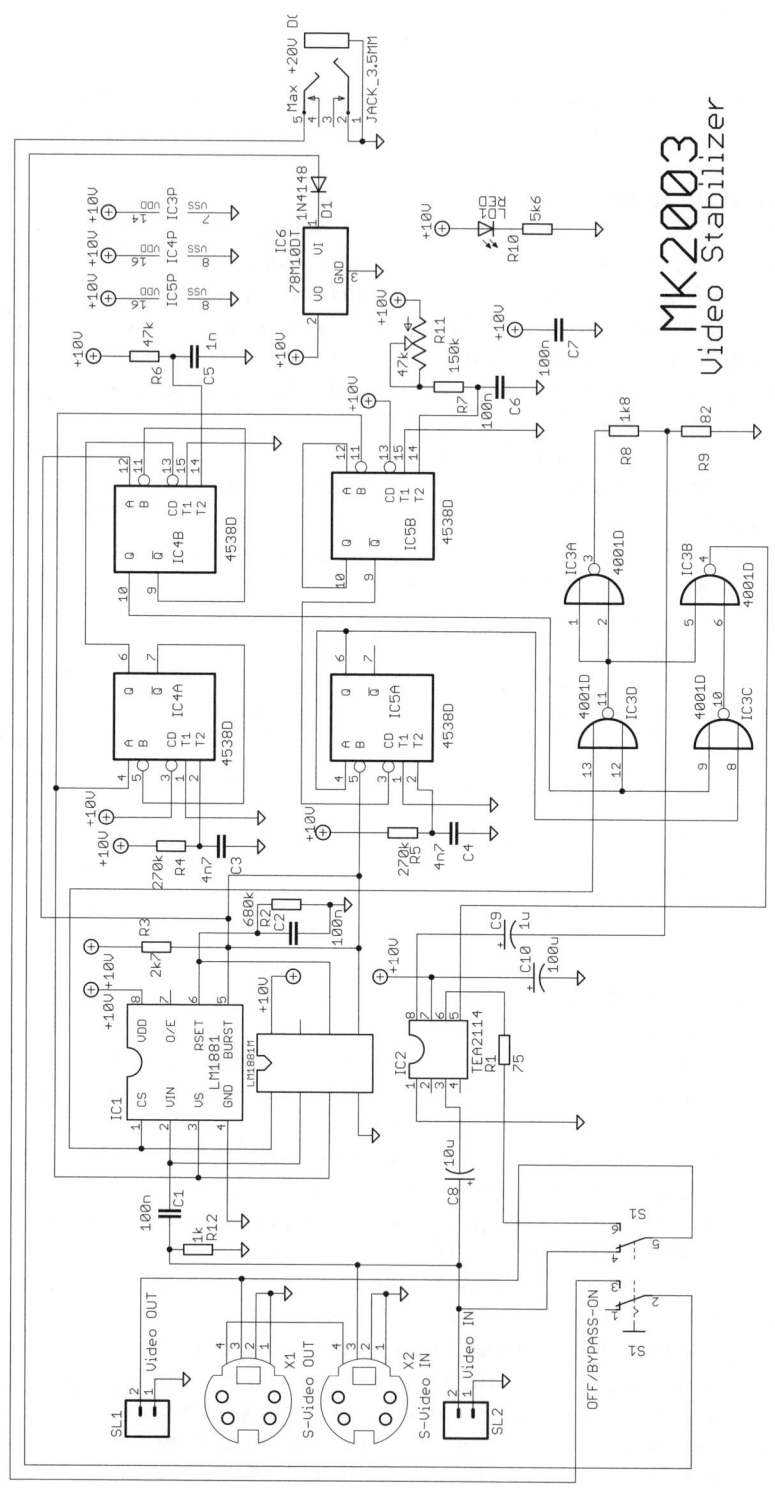

Fig. 37

7.5 Analogue Anticopy Protection

logic level '0' by the output Q of IC4A.

At the Q output of IC4B, to coincide with the BURST's rising edge coming from pin 5 of IC1 and connected with the input 'A' of IC4B (pin 12), there will be a positive pulse of about 48 microseconds, a value calculated from R6 & C5 timing.

To eliminate EOF pulses we need a vertical sync whose rising edge (from the pin 3 of IC1) will guide the monostable IC5B to produce a pulse of about 18ms at its Q\ output (pin 9), corresponding roughly to a field timing, and finely adjustable with R11. This unlocks the monostable IC5A to produce, at the first falling edge of the BURST signal, a pulse of about 1.2 ms, available from the output Q of IC5A (pin 6).

The signals from IC4B/pin 10, IC5A/pin 6 and IC1/pin 1 (i.e. CS = Composite Sync), combined by NOR logic gates IC3A/B/C/D, finally produce the two signals that eliminate the anticopy protection.

More precisely on IC3A/pin 3 are the composite sync pulses 'cleaned' by EOF pulses, while on IC3B/pin 4 is the cancellation signal to the dancing pulses.

7.5.4 The output stage & Power Supply

IC2 (TEA2114) is responsible for the 'analogue' elimination of the anticopy signal. In fact it is a video switch with 2 inputs and its output can drive a 150Ω load. The inputs are clamped and the buffer amplifier has 18MHz bandwidth.

The two inputs IC2/pin 1 and IC2/pin 8, respectively, are fed from the video input (through C8, a non-polarized capacitor) and the cleaned signal of composite sync correctly attenuated by a resistive network formed by R8 & R9 via C9. IC2's control switching pin (i.e. pin 5) is fed by the signal coming from the output of the NOR IC3B, which has a CMOS level signal sufficient to drive it.

IC2's output (pin 6) through R1 and S1 is passed to the composite video and S-Video mini-DIN output connectors. Also S1 makes a by-pass and an on-off switch for the circuit. IC6 is a SMD 10V voltage stabilizer and provides a +10V supply voltage necessary for operating all ICs correctly. A diode in series with its pin VI (=Voltage Input, IC6/pin 1) ensures the correct polarity which is then confirmed by illuminating LED LD1. Except for both S-Video mini-DIN sockets and the plug supply connector, which are placed on the PCB to simplify the construction, the other connectors (S1 and the two RCA video plugs) must be placed externally on a small case (see below).

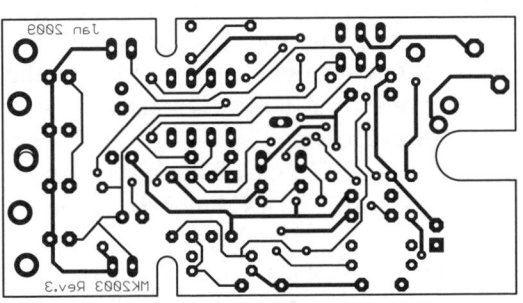

Fig. 38 - MK2003 Video Stabilizer's PCB bottom layer

7.5.5 Assembling & Soldering

Having already gained access to the double-sided PCB, the arrangement and soldering of the small amount of components should not give particular difficulties. However always pay close attention to orientation of all the polarized components and the component silkscreen layout should resolve all doubts about the position in which the few components are positioned.

Although generally it is always better to solder first all the discreet and passive components (resistors, capacitors, diodes, etc.), in this case it is preferable to proceed first with the soldering

Video recorder systems

Fig. 39 - MK2000 Video Stabilizer's PCB

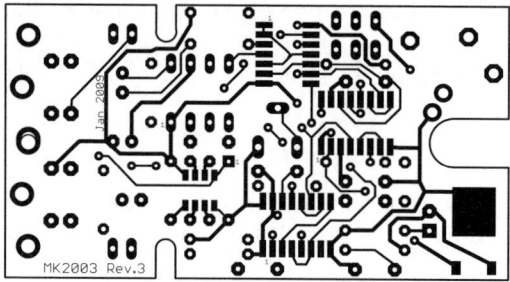

Fig. 40 - MK2003 Video Stabilizer's PCB top layer

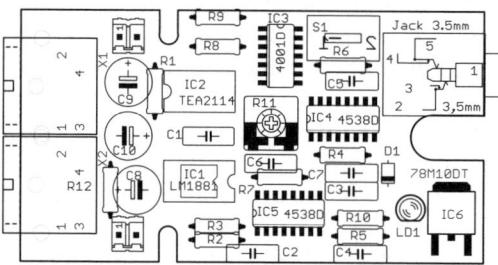

Fig. 41 - MK2003 Video Stabilizer's component layout

of SMD devices and the two 8-pin ICs. This is because SMDs are very small and they need lots of precision attention for soldering without sockets. A good 1-mm tip soldering iron can help make a clean job, paying attention to the SMD package's reference point which sometimes is not visible easily on these products or even not present! The better way is to solder only one pin placed in a corner, then align the device to match the pads and finally solder the remaining pins.

The next step is to solder IC1 & IC2 preferably without sockets to keep a low profile PCB in case we want to insert the circuit into a small box. If convenient, IC1 can be substituted with the SMD version (LM1881M) since the PCB is already prepared to solder it. Then it is the turn of the resistors and capacitors (warning: C9 & C10 are polarized!), the diodes D1 & LD1 and the S-Video & power supply connectors (Fig. 38, 39, 40, 41).

7.5.6 Calibration

If everything was assembled correctly next visually inspect, even with a magnifying glass, to avoid any short circuits between the tracks or SMD IC pins. Connect a DC voltage of +13...20VDC to the power jack connector. LED LD1 should turn on immediately, otherwise control the voltage polarity through a multimeter's probes connected between pin 1 (VI) of IC6 and ground (+12...20VDC) or between pin 2 (VO) and ground (+10VDC, with a ±5% of tolerance). If there is a suitable voltage then probably LD1 is faulty or its polarity is reversed. Simple to fix!

An initial good sign is if a video signal connected to the RCA plug input played from a video source like a DVD player is visible even to just some extent on a TV monitor connected to the RCA plug output.

There are two methods for calibrating the only trimmer present in this circuit: empirical and instrumental. The former is quite easy to do but inaccurate, the latter needs an oscilloscope.

The empirical method consist of judging the picture directly on the screen: when it is stable the trimmer is calibrated. A horizontal black bar at the bottom of the screen (its height can be regulated by R11) proves that the EOF pulses are eliminated.

Using an oscilloscope makes calibration much simpler. Detect the EOF pulses on the oscilloscope's screen and rotate the trimmer until they are suppressed.

If there are still problems, they could be caused by excessive tolerances of R6 & C5: it

7.5 Analogue Anticopy Protection

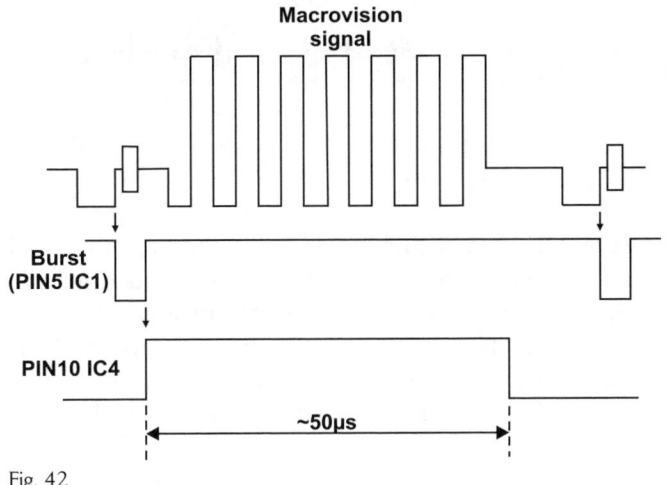

Fig. 42

could be so high or low that the resultant timing does not allow full coverage of 'dancing' pulses or even it could cause missing syncs. If that happens, replace R6 with a 100kΩ trimmer and adjust to obtain from pin 10 of IC4B a 50µs positive pulse. A good idea could also be to substitute empirically (i.e. 'select on test') with a nearest value resistor to obtain the same effect, assisted by oscilloscope.

In any case the included oscillograms can help to identify any trouble (Fig. 42 & Fig. 43).

7.5.7 Inside a box

A small plastic box to house the circuit is preferable to a large metal one, first of all because the former is easier to cut or profile to match the sockets soldered on MK2003's PCB. The switch S1, the power supply LED LD1 and composite video input and output RCA (or BNC if you prefer) sockets have to be fixed externally and their terminals have to be connected to

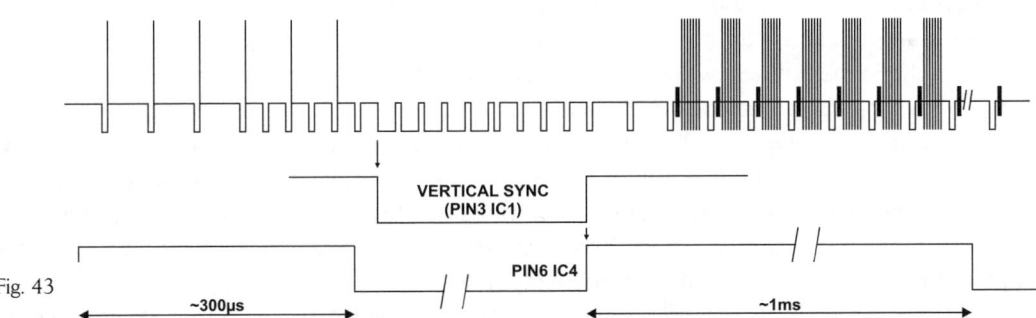

Fig. 43

MK2003 via some insulated but non-shielded wires even for video signal since this last has a low impedance. If you want you can incorporate the previous video Fader/Enhancer/Distributor circuit inside the same box, using the MK2003 in video composite mode only but inserted immediately after the player's video output. The MK2003' video output will feed the Fader/Enhancer which, consequently, will furnish a video signal sufficient to drive three recorders simultaneously.

Fig. 44

7.5.8 Bad thoughts...

Once disturbing pulses are removed by MK2003, it could even become possible to record the video signal (Fig. 44), but we want to highlight that this project is not intended for this and such action is discouraged. It is generally illegal to record copyrighted material otherwise there are heavy penalties. It is therefore our duty to warn you for the specific purpose of this project even though, according to the various legislatures of each country, in principle, making a copy for personal use is permitted as long as you own the original and this copy does not leave home.

The people who developed the copy protection standards are the first to admit they cannot stop well-equipped pirates from fighting the illegal video recording preventative measures, but it is always better to refrain!

7.6 Some conclusive considerations

Although the necessity and importance of recording audios and videos on tape or cassette is nowadays dramatically reduced, it is clear that it is a distant future when the age of tape will disappear into the sunset. It has been suggested that the end of tape will be within a few years based on the increasing importance and acceptance of new methods for recording on optical medium, hard disk and solid-state memory. However, the 'old' tape-based storage medium has found a niche in the business world, by lovers of nostalgia, analogue-signal amateurs and professionals, including also some consumer users.

In addition, videotape is still terribly efficient. A 1-hour digital video tape in the MiniDV cassette format can store up to 15GB of audio-video data occupying an area of 65x45x10 mm! Whole shelves containing several thousands of mini cassettes can hold hundreds of terabytes of data which inter-alia provides a green way to store data in competition perhaps with the most modern methods that waste kilowatts of energy power.

Moreover a tape is still considered more reliable than solid-state supports. If a part of a tape is broken or damaged, it can be rejoined or eliminated losing only the data of that part. If a solid-state medium is mechanically broken or electronically faulted, there is no simple way to recover data! On the other hand a strong magnetic field can totally erase tape, an event that a solid state memory and an optical disc cannot be affected. However CDs and DVDs are vulnerable to scratches, excessive heat or chill. Every innovation in storage medium generally improves features of previous supports but at a certain cost.

Fifty years have passed since that obscure object called the 'Quadruplex Video Recorder' which, for the first time, allowed broadcasters to record and relay events and news without the anguish of live transmissions. It is a privilege that end users have enjoyed successively with

7.6 Some conclusive considerations

the chance to choose when and how to see a movie or a TV programme previously recorded without keeping them glued to the armchairs of their living rooms.

Nowadays we almost completely use the hard drive and DVD recorders. A long way has really gone from those days, some fifty years ago, when that Quadruplex half-ton machine hummed with its rotating massive heads while etching the tape!

Nevertheless, since the arrival of video tape recording technology, it has been possible to capture countless happy and tragic events that have made history, often on a whim, like the last pictures of Princess Diana impressed in a surveillance recorder system before the fatal accident. The way to hand down events to posterity has changed forever.

With this chapter I think that it is important to remark how incredibly much research, development and progress there has been behind the soft push with which we can now lazily touch the REC button on our recorders.

To be continued...

The growth of television is not only strictly bound to technological developments and innovations over years. Just consider the differences between 'an invention' and, less easy to achieve, 'an innovation'. The former was generally concocted by an individual inventor often based on a fortuitous discovery or on experimenting, something that requires creativity and imagination and often without any specific good grounding. The latter, contrarily, was (and still is) achieved by a crew of industrial researchers financed usually by private multinational companies or by a national government in order to specify common standards for each new invented or innovated device in a context of an economic and marketing system where producers and users meet the same exigencies.

Incidentally, the elevated number of would-be scientists and talented inventors of equipments and devices, some of whom are dead or destitute, has demonstrated the need for a targeted planning during the innovation phase inside a specific technological field.

So, in the first part of the trip inside the television world, many names are tied to television evolution often in mutual competition. Later they were absorbed into the brand of great companies or industries whose CEOs generally took the credit for the introduction of a new device, equipment or gadget, or for a new discovery, innovation or contrivance, behind which there was always a very skilled team composed by many engineers, technicians, chemists, etc., who did the real work!

Through this book we have talked about how such names and companies have worked in achieving outcomes in the television field travelling from the first black & white mechanical television experiments to fully electronic colour transmission across several television inventions such as a multitude of picture, pickup and recording devices.

Based on those researches and developments we have learned what synchronisms are and how to generate a suitable composite video signal, in addition to other information. We have built by ourselves a sync pulse generator (the *SPG625*), a sophisticated video instrument (the *PAL Colour Bars Generator*), a funny experimental solid-state oscilloscope (the *LED Oscilloscope*), a circuit which will improve the 1st copy of an analogue video recording (the *Video Enhancer & Distributor*) and a device which cleans up protected movies (the *MK2003 Video Stabilizer*).

However the fascinating voyage around the video world is not finished here. The subject is so huge that this work requires to be split into further parts.

The first part of this work (i.e. this book) is concerned with fully analogue aspects of video signals, in addition to historical hints, while the second part (next book) will describe how the incoming new electronic analogue and digital technologies allow us to mix, cut, overlay, modify or elaborate the analogue video signal. Also we will see how to add some digital information over an analogue video signal without degradation or how it is possible to freeze, squeeze, compress, expand, magnify, rotate, flip, etc., a video image sequence in real time.

All the next arguments are as interesting as unfamiliar and intriguing: **STAY TUNED!**

Bibliography

In order to write a book like this you must rely not just on personal experience over the years, but also on the work of other authors who have prior researched by collecting letters and reports, oral histories and memoirs of participants, excerpts from specialized publications, newspapers and documents. Even the internet has been invaluable and references to web pages that contain topics discussed and are no longer traceable on the internet may be found out by entering the URL into the Wayback Machine (Internet Archives) at http://web.archive.org.

Books

Bali, S.P. *Colour Television: Theory and Practice*. New Delhi (India): Tata McGraw-Hill, 1994.

Daniel, Eric D., and C. Denis Mee and Mark H. Clark. *Magnetic recording: the first 100 years*. Piscataway, NJ (USA): IEEE Press Editorial Board, 1999.

Gulati, R.R. . *Monochrome and Colour Television - Revised Second Edition*. New Delhi (India): New Age International (P) Ltd, Publishers, 2005.

Gupta, R G. *Television Engineering and Video Systems*. New Delhi (India): Tata McGraw-Hill, 2006.

Howett, Dicky. *Television innovations: 50 technological developments*. Tiverton, Devon (UK): Kelly Publications, 2006.

Ibrahim, K. F. . *Newnes guide to television and video technology*. Oxford(UK): Elsevier, 2007.

Keith, Jack. *Video demystified: a handbook for the digital engineer - Fourth Edition*. Burlington, MA (USA): Elsevier, 2005.

Magoun, Alexander . *Television : the life story of a technology*. Westport, Connecticut (USA): Greenwood, 2007.

Robin, Michael, and Michel Poulin. *Digital television fundamentals*. New York (USA): McGraw-Hill, 2000.

Terenzi, Giorgio, and Giuseppe Commissari. *Il nuovo videolibro*. Bologna (Italy): Ulrico Hoepli Editore SPA, 1998.

Todorovic, Aleksandar Louis. *Television technology demystified: a non-technical guide*. Burlington, MA (USA): Elsevier, 2006.

Trundle, Eugene. *Newnes guide to television and video technology*. Oxford(UK): Elsevier, 2001.

Watkinsons, John. *Television fundamentals*. Oxford(UK): Focal Press, 1994.

Weise, Marcus, and Diana Weynand. *How video works*. Burlington, MA (USA): Elsevier, 2007.

Whitaker, Jerry, and Blair Benson. *Standard Handbook of Video and Television Engineering - Fourth edition*. USA: McGraw-Hill, 2004.

Wood, John L., and Trevor Brown. *Amateur Television Handbook*. Thame, Oxon (UK): BATC, 2000.

Magazines & Publications

Gatti, Marco. *"Due novità elettroniche rivoluzionano la musica e la televisione"*. Rome (Italy): Scienza e Vita. Maggio 1955.

Bibliography

Gatti, Marco. *"La Televisione a spirale"*. Rome (Italy): Scienza e Vita. Febbraio 1955.

Pirastu, Massimo. *"Se ne vedono di tutti i colori – Solo un'analogia tra l'occhio e la macchina fotografica"*. Milan (Italy): Rusconi Editori Associati. Scienza e Vita Nuova, Novembre 1980.

Valva, Giuseppe D'Ayala. *"A che cosa serve la Televisione?"*. Rome (Italy): Scienza e Vita. Giugno 1952.

Valva, Giuseppe D'Ayala. *"Che cosa è la televisione? I tre momenti della trasmissione e ricezione"*. Rome (Italy): Scienza e Vita. Aprile 1952.

Valva, Giuseppe D'Ayala. *"Conclusa la fase sperimentale la Televisione Italiana è pronta ad affrontare il suo impegnativo esame di maturità"*. Roma (Italy): Scienza e Vita. Ottobre 1953.

Valva, Giuseppe D'Ayala. *"I principali sistemi allo studio per la televisione a colori"*. Roma (Italy): Scienza e Vita. Aprile 1950.

Valva, Giuseppe D'Ayala. *"La Televisione nel mondo"*. Roma (Italy): Scienza e Vita. Novembre 1953.

_____. *Scuola Pratica di Video*. Sesto S. Giovanni, MI (Italy): Alberto Peruzzo Editore, 1984

Web Pages

Barnett, Alan, and Martin Evans. *"PALsite: The home of the PAL video system"* (http://www.palsite.com), visited June, 2009.

Beale, John. *"New To Video - Information for Getting Started"* (http://www.bealecorner.com/trv900 /new2video.html), visited June, 2009 and *"Resolution Test Patterns"* (http://www.bealecorner.com /trv900/respat/), visited July, 2009.

Carter, R. L. *"DigiCam History Dot Com"* (http://www.digicamhistory.com), visited September, 2009.

Christiansen, P., and G.R. Kohlbacher. *"CCD for PAL decoder"*, Solid-State Circuits, IEEE Journal of Volume 16, Issue 3, Jun 1981 (http://ieeexplore.ieee.org/xpl/freeabs_all.jsp?arnumber=1051562), visited June 2009.

Donnelly, David F. The Museum of Broadcast Communications (MBC), *"Color Television"* (http://www.museum.tv/archives/etv/C/htmlC/colortelevis/colortelevis.htm), visited April, 2009.

Elen, Richard G. *"Baird's independent television"* (http://www.transdiffusion.org/emc/baird/baird_itv.php), visited January, 2009.

Goldwasser, Samuel M. *"TV and Monitor CRT (Picture Tube) Information"* (http://www.walshcomptech.com/repairfaq/REPAIR/ F_crtfaq.html), visited February, 2009.

Heap, Steve. *"FML Test Card Maker"* (http://www.oodletuz.fsnet.co.uk/soft/tcmaker.htm), visited May, 2009.

Iisakkila, Mika. *"Video recording formats"* (http://users.tkk.fi/iisakkil/videoformats.html), visited October, 2009.

Keith, Jack. Intersil – *"Consumer Analog RGB and YUV Video Formats"* (http://www.intersil.com /data/an/an9727.pdf), visited May, 2009.

King, Bevis. *"Worldwide TV Standards - A Web Guide"* (http://www.ee.surrey.ac.uk/Contrib/WorldTV), visited March, 2009.

Kittelsen, Steve. *"Info on the three color TV systems"* (http://www.nmia.com/~roberts/vidstd), visited April, 2009.

Lapini, Gian Luca. Storia di Milano - *"Milano e la televisione"* (http://www.storiadimilano.it/

citta /milanotecnica/televisione/tv.htm), visited February, 2009.

Lee, Dana M. *"Television Technical Theory: Unplugged - Version 5.0"* (http://www.danalee.ca/ttt), visited May, 2009 and September, 2009.

Mannu, Sergio. *"La tv a colori in Italia"* (http://www.pagine70.com/vmnews/wmview.php?ArtID=650), visited February, 2009.

Mayes, Lawrence. *"TV Synchronisation"* (http://freespace.virgin.net/ljmayes.mal/var/tvsync.htm), visited February, 2009.

McCormick, Colin. *"The Beta (Betamax) Video Tape Format in the UK"* (http://www.colin99.co.uk /beta.html), visited September, 2009.

McLean, Don. *"The World's Earliest Television Recordings"* (http://www.tvdawn.com), visited September, 2009.

Mitchell, Frank. *"A Very Concise History of Test Cards"* (http://www.testcardcircle.org.uk/tchistory.html), visited May, 2009.

Pellacani, Rolando. *"Le telecamere Broadcast"* (http://www.dreamvideo.it/video/ tlc_broadcast_sommario.htm), visited August, 2009.

Pemberton, Alan. *"Pembers' Ponderings"* (http://www.pembers.freeserve.co.uk), visited July, 2009.

Peters, Jean-Jacques. *"A History of Television"* (http://www.etsu.edu/cas/COMM/broadcasting/broadcastingweb/tvhistory/dvb_tv-history.htm), visited March, 2009.

Powell, Evan. *"So what is "Component Video" anyway?"* (http://www.projectorcentral.com /component.htm), visited February, 2009.

Poynton, Charles. *"Color FAQ - Frequently Asked Questions Color"* (http://www.poynton.com /notes/colour_and_gamma/ColorFAQ.html), visited April, 2009.

Robin, Michael. *"The color bars puzzle"* (http://broadcastengineering.com/mag/ broadcasting_color_bars_puzzle), visited March, 2009.

Runyon, Steve. The Museum of Broadcast Communications (MBC), *"Television Technology"* (http://www.museum.tv/archives/etv/T/htmlT/televisionte/televisionte.htm), visited Febrary, 2009.

Scheida, Wolfgang. *"45 Years Anniversary of Walter Bruch's PAL Color Television"* (http://www.radiomuseum.org/forum/45_years_anniversary_of_walter_bruchs_pal_color_television.html), visited March, 2009.

Trexler, Kris. *"The Color Television Revolution"* (http://www.ev1.pair.com/colorTV), visited March, 2009.

Yanczer, Peter F. *"Telorama - Television For The Experimenter"* (http://www.televisionexperimenters.com /home.html), visited June, 2009.

Baird Television (http://www.bairdtelevision.com), visited January, 2009.

Comitato Guglielmo Marconi International – Televisione Marconi. *"La TV nasce nel 1926, in un grande magazzino..."* (http://www.radiomarconi.com/marconi/televisione.html), visited May, 2009.

Commodore: 2080 (http://www.amiga-hardware.com/showhardware.cgi?HARDID=864), visited April, 2009.

Early Television Museum (http://www.earlytelevision.org/), visited February, 2009.

Improvements in apparatus for the remote transmission of pictures - GB Patent 750187 (http://www.wikipatents.com/gb/750187.html), visited March, 2009.

Bibliography

"ITT Nokia Tv Color Digivision (1989)" (http://www.torinointernational.com/spot80/spot/itt-nokia-tv-color-digivision-1989.html), visited June, 2009.

LCD TV Reviews - *"History of Television"* (http://www.lcd-tv-reviews.com/_history_of_television.php), visited January, 2009.

MHP – The Test Card Gallery – *"A Little Bit of History"* (http://www.meldrum.co.uk/mhp/testcard/history.html), visited April, 2009 and *"Television Graphics Around the World"* (http://www.meldrum.co.uk/mhp/testcard/around_world.html), visited May, 2009.

Narragansett Imaging - *"Camera Tubes – Background Information"* (http://www.nimaging.com/products/tubes/background.html), visited July, 2009.

Narrow-bandwidth Television Association (http://www.nbtv.org), visited April, 2009.

Television History - The First 75 Years (http://www.tvhistory.tv), visited January, 2008.

The Cathode Ray Tube site - Electronic glassware - History and Physics Instruments (http://members.chello.nl/h.dijkstra19), visited July, 2009.

The Early Video Project (http://www.davidsonsfiles.org), visited April, 2009.

The Experimental Television Center's Video History Project (http://www.experimentaltv-center.org/history), visited April, 2009.

This is SPTV (http://www.sptv.demon.co.uk), visited April, 2009.

Video chips and circuits (http://icc.skku.ac.kr/~won/electro/videochips.html), visited February, 2009.

Video Interchange - Vintage Audio - Video Transfer Recording Rerecording Conversion Services Recovery - Restoration - Remastering - Fire - Flood - Mold - Antique & Obsolete Formats (http://www.videointerchange.com), visited August, 2009.

Video Preservation Website (http://videopreservation.conservation-us.org), visited March, 2009.

Wikipedia (http://en.wikipedia.org/wiki), visited March, April, May, June, July, August and September, 2009.

Special thanks to:

Mr. Giuseppe Musumeci, photographer and cameraman, who helped me in writing the chapter about video cameras;

Prof. Dana M. Lee of Ryerson University in Toronto (Canada) for his immense helpfulness to use anything within his work *"Television Technical Theory: Unplugged - Version 5.0"*, online version (http://www.danalee.ca/ttt);

Mr. Paul J. Ridgway for his invaluable helpfulness in writing and reviewing this work.

Index

0-9
16:9, 28
1-H delay line, 81
2006 Consumer Electronics Show, 147
240-lines standard, 22
2D Adaptive comb filter, 98
2-Line comb filter, 98
2-tube cameras, 162
3D Motion Adaptive comb filter, 98
3-Line comb filter, 98
3-tube cameras, 162, 163
4:3, 27, 28
405-lines standard, 22, 23, 101
4-tube cameras, 161
625-lines standard, 22
74ACT715, 35
819-lines standard, 22

A
A2080 Monitor, 137
A2300/A2301 Amiga Genlock, 75
A520 Amiga Modulator, 68
Aben, Herman J.S., 114
Academy of Motion Picture Arts and Sciences, 27
Accomodation, 46
Acorn Electron, 30
Active Matrix Liquid Crystal Display (AMLCD), 145
Active Pixel Sensor (APS), 169
AD722, 75
AD723, 75
AD724, 75, 102
AD725, 75
Additive Mixing, 54, 161
ADV717x, 67
ADV73xx, 67
ADV7802, 98
Allen, Clile C., 170
Amiga 1000, 68
Amiga 1200, 73
Amiga Genlock, 7
Analogue Anticopy Protection, 191
Analogue Encoders, 75
Analogue Decoders, 81

205

Index

Analogue input encoder chips, 68
Analogue Protection System (APS), 191
Angenieux, Pierre, 171
Anode, 125
Anti Comet Tail (ACT), 164
Aperture Grille, 140
Apple Tube, 143
Aqueous humour, 46
Aspect Ratio, 27
AT & T, 18
Atari 800, 58
Audio Compact Cassette, 180
Audio Level and Input Controls, 172
Auto Chroma Control (ACC)circuit, 78
Auto Iris, 170
Auto White/Black Balance, 171
Autofocus, 170
Automatic Gain Control (ACG), 190, 191
Automatic Tracking Circuit, 181
Auto-Tint circuit, 59
AX tubes, 141
Azimuth, 177, 181

B

BA7046, 44
Back & Front Porches, 24
Baird Television, 134
Baird Undersock, 16
Baird, John Logie, 8, 9, 10, 16, 17, 22, 126, 134, 142, 156, 175
Balanced Modulators, 55
Bars & Raster Generator, 103
Bars & Tone Generator, 108
Bayer Filter, 166
BBC Microcomputer, 30
BBC Tuning Signal, 101
BBC, 22
BBC2, 23
Beam-Index Tube, 143
Bell Labs, 18
Betacam SX, 178
Betacam, 172, 180, 182, 184, 189
Betamax, 178, 179, 180, 181, 183, 184
Betamovie, 179
BH7236, 73
Bias, 187

Index

Bing Crosby Enterprises (BCE), 175
Bitzer, Donald L., 146
Black & Burst, 70, 101
Black Box, 187
Black Level, 24, 26, 27, 40, 41, 42, 53, 60, 68, 75, 79, 106, 107, 130, 172, 190, 192
Blind spot, 47
Blue-ray Disc, 182
Blue-ray player, 183
Boyle, William, 165
Braun Tube, 126, 135
Braun, Karl Ferdinand, 16, 125, 135
BrightSide Technologies, 146
Broad Pulses, 24, 26
Brody, T. Peter, 145
Bruch, Walther, 17, 20, 99
BU1424K, 67
Bucket Brigade Device (BBD), 165
BVU (Broadcast Video U-matic), 178
BVU Superior Performance (BVU-SP), 178

C

Cairns and Morrison Ltd., 175
Calibration Aider, 119
CalMAN HTPC Pattern Generator, 119
Camarena, Guillermo Gonzales, 18
Camcorder, 108, 152, 172, 179, 181, 182, 183, 184
Cameraman, 170, 172, 178
Campioni, Armando, 20
Capsule Endoscopy, 172
Carey, George, 15
Caselli, Abbe Giovanni, 15
Cathode Beam Modulation, 138
Cathode Ray Tube (CRT), 16, 126, 135
Cathode Ray, 16
Cathode, 125, 157
CBS (Columbia Broadcasting System), 17
CCD (Charge Coupled Device), 87, 156, 160, 165, 166, 167
CCIR (Comite Consultatif International pour la Radio), 22
CCIR625, 22, 23, 29, 35, 39, 101
CD22402, 33
Central fovea, 47
Chalnicon, 160
Chessboard raster, 108
Chromatron, 140
Chrominance, 18, 55, 56, 62, 64, 75, 78, 87, 97, 133, 188

Index

Chromoscopic Adapter, 18
Clean PAL, 64
CMOS image sensors, 169
C-Mount, 171
Cold White Colour, 171
Colour Bar Generator, 104, 133
Colour Burst, 57, 66, 78, 110, 191, 192
Colour decoders chips, 81
Colour Difference Signals, 54
Colour Encoders, 65
Colour Field Sequential System, 48
Colour Killer, 78
Colour Standard Systems, 46
Colour Temperature, 171
Colour Transient Improvement circuit (CTI), 86
Comb filter, 78, 79
Commodore 58, 68
Compact Disc, 181
Compact Flash (CF), 183
Compagnie Francaise de Television, 19
Complementary Colours, 54
Compressed Time Division Multiplexing (CTDM), 182
Control Camera Unit (CCU), 162, 172
Control Grid, 125
Convergence, 139
Copyguard, 191
Cornea, 46
CPS Emitrons, 158, 159
Crescent Wood Road Laboratory, 142
Crookes, Sir William, 16, 125
Cross Colour, 58, 97
Cross Luminance, 58
Crosshatch raster, 108
CSTN=Color Super Twist Nematic, 145
CTI (Colour Television Incorporated), 51
CVBS, 58, 86, 97, 98, 104
CXA1145, 73
CXA1213S, 90
CXA1214P, 90
CXA1645, 73
CXA2075, 73
CXD1159AQ, 35
CXD1217M, 35

CXD1254AR, 35
CXD1257AR, 35
CXD1261AR, 35

D
D2-MAC, 64, 182
D8, 181
D9, 182
DAI Personal Computer, 66
Dark Current, 165
Day, William, 134
De Forest, Lee, 125
De France, Henri, 19, 59
De Gaulle, President, 19
Decoder, 50
Definition, 164
Deflection Yokes, 126, 137
Degaussing Coil, 140
Delay Line, 20, 57, 61, 69, 75, 79, 80, 81, 82, 86, 89, 100, 166, 186
Delta-Gun Shadow Mask kinescope, 140
den Bak, Peter, 118
Detail Level, 171
Dichroic mirrors, 49, 161, 162, 163, 171
Dieckmann, Max, 157
Differential Phase Error, 57
Digit2000, 94
Digital & Pseudo-Digital Decoders, 91
Digital 8, 181
Digital Betacam, 182
Digital inputs encoder chips, 65
Digital Micromirror Device (DMD), 151
Digital Pixel Sensor (DPS), 169
Digital Versatile Disk (DVD), 182
Digital VHS (D-VHS), 182
Digital video recorders, 172, 174
Digital Video, 13, 14, 117, 181, 182, 189
Digital-S, 182
Digivision, 94, 96
Display devices, 126
DLP (Digital Light Processing) Projectors, 151
DNS (Dynamic Noise Suppression), 181
DOC (Dropout Compensator), 189
Dots, 108

Index

DSTN=Double Layer Supertwist Nematic, 145
Dual Head, 153
DVCAM, 182
DVD Recorders, 182
DVDPRO, 182
DVD-R, DVD+R, 182
DVD-RW, DVD+RW, 182
Dynamic Convergence, 140
Dynamic Track Following (DTF), 180
Dynode, 159

E

EBU (European Broadcasting Union), 104
Edison effect, 125
Edison, Thomas Alva, 15, 124
Eidophor, 150, 151
EL4581, 42
Electric lamp, 124, 125
Electronic Field Production (EFP), 162
Electronic News Gathering (ENG), 162
Encoder, 49
Equalisation Pulses, 25, 26
ETP-1 monoscope, 115
Euroconnector, 96
EV-200, 177
Even Field, 26
Exmor-R, 169
Extended Definition Beta (ED-Beta), 180
Extended PAL, 64
Extended Playing (EP), 180
Extra High Tension (EHT), 137
Eye lens, 46
Eye, Human, 8, 10, 14, 20, 25, 46, 47, 49, 52, 58, 59, 61, 62, 124, 133, 150, 156, 157, 159, 170, 171, 172

F

Faggin, Federico, 169, 173
Faraday, Michael, 15
Farnsworth, Philo Taylor, 16, 157
FBAS, 58
FCC (Federal Communications Commission), 18, 26
Ferranti, 30
Field Emission Display (FED), 143
Film Recorder, 175
Firewire, 183

Fischer, Fritz, 150
Fleming Oscillation Valve, 125
Fleming, John Ambrose, 125
Flyback Transformer, 137
Focus, 46
Footage Transferring, 189
Foveon X3, 169
Frame Interline Transfer CCD, 167, 168
Frame Transfer CCD, 166, 167
FuBK monoscope, 115

G
Gain Switch, 171
Geißler, Johann Heinrich Wilhelm, 125
Genlock, 7, 75, 93, 162
Goldberg, Al, 104
Goldmark, Peter Carl, 48
Goldsmith, Alfred Norton, 135
Goldstein, Eugen, 14, 125
Grassmann, Hermann, 52, 53
Gretener, Edgar, 150
Grid Beam Modulation, 138
Grid raster, 108
GS1881, 43
GS4881, 43
GS4882, 43
GS4981, 43
GS4982, 43
Guthrie, Frederick, 125

H
Hannover Bars, 57
Hannover Blinds, 57
Hard Disk Drive Recorder, 183
Hazeltine Corporation, 51, 52
HBLK, 24
HD DVD, 182
HD44007A, 35
HDCAM, 182
HDD Recorder, 183
HDMI (High-Definition Multimedia Interface), 122, 155
HDR-TV Display, 146
Heap, Steve, 120
Heilmeier, George Harry, 145
Helical Scan, 177, 182, 183, 188

Index

Hell, Rudolf, 157
Henry, Joseph, 15
Heptode, 125
Hersee, Carole, 115, 123
Hersee, George, 115
Hexodes, 125
High Dynamic Range (HDR), 146
High-Band, 178
Hi-Sensicon, 160
Hittorf, Johann Wilhelm, 15, 125
Holmes, David D., 104
Holodeck, from Star Trek serials, 155
Home Theatre System, 151
Horizontal Blanking, 24
Horizontal Deflection Electrodes, 128
Horizontal Sync, 24
HS, 24
HTPS (High-Temperature Polysilicon), 153
Hue, 50, 57, 61
Human Interface Technology Lab, 150
Hysteresis, 187

I

I2C, 93
Iconoscope, 16, 157, 164
IEEE 1394, 183
iFire Technology Inc., 148
Image Dissector, 16, 157, 159, 164
Image Iconoscope, 158
Image Orthicons, 159, 161, 163, 164
Image Time Persistence, 164
Imaging Associates, 119
IMLED (Individually Modulated Array of LED), 146
Impact Vision 26, 70
In Phase, 54
Indesit, 20
Indextron, 143
Indian Head Monoscope, 115
In-Line Shadow Mask, 140
Interlace 2:1, 25
Interline Storage Registers, 168
Interline Transfer CCD, 168
Iodopsin, 47
IRE, 130
Iris, 170

ISA (Identificazione a Soppressione Alternata), 20, 64
ITT International, 94
ITV, 23
Ives, Herbert Eugene, 18

J
Jenkins Television Corporation, 17
Jenkins, Charles Francis, 17

K
KA2195D, 73
KA2198BD, 73
Kell Factor, 28
Kell, Bedford & Trainer, 28
Kenotron, 125
Kine, Kinescope Recording, 175
Kinescope, 17, 18, 48, 50, 112, 114, 124, 135, 140, 141, 158, 175
Kinescope, Tri-Colour Shadow Mask, 19, 50

L
La Spina, Angelo, 7
Laboratoires R. Derveaux, 44
Large screen displays, 150
Larky, David, 104
Laser Video Projector, 152
Lawrence, Ernest, 140
LC78011E, 67
LCD Projectors, 151, 152
LCOS (Liquid Crystal On Silicon), 153
Le Prince, Louis Augustin, 15, 45
LED Matrix, 132
LED Oscilloscope, 131
LED Videowall, LEDwall, 153
LED, 154
Leddicon, 160
Lehmann, Otto, 144
LEP (Light Emission Polymers), 147
Line Blanking, 24
Line Output Transformer, 137
Line Sync, 24
Linear CCD, 165
Linear NIR, 64
Liquid Crystal Display (LCD), 144, 145, 146, 147, 148, 150, 154
LM1881, 40, 41, 42, 43
LM1882, 35

Index

LM1886, 67
LM1889, 66
LMH1251, 98
LMH1980, 43
LMH1981, 43
Long Playing (LP), 180
Luminance, 18, 50, 53, 54, 56, 58, 59, 60, 61, 62, 64, 67, 69, 73, 75, 76, 78, 79, 80, 86, 87, 89,
97, 104, 106, 130, 146, 162, 163, 182, 188, 193

M

MAA20xx, 94
MAC (Multiplexed Analogue Components), 64
Macrovision, 91, 191, 192, 193
Magnetic Deflection Yokes, 128
Magnetic Tape, 12, 13, 63, 174, 176, 177, 187
Marconi, Guglielmo, 15, 126, 136
Marconi-EMI, 22, 101
MB3516A, 73
MC13077, 70
MC1377P, 68
MC1378, 70
MC1496, 65
MC1596, 65
MC44145, 44
MDDP (Micro Device Display), 152
Megapixels, 166
Meucci, Antonio, 15, 45
Micro Device Display Consortium (MDDPC), 152
Micro Device Display Projection Televisions (MDDP), 152
MII, 182
MiniDV, 181, 184, 198
MiniDVD-R(W), 182
Minority Report, Steven Spielberg's, 155
MK2003 Video Stabilizer, 193, 195, 196, 197, 200
ML87V21071, 98
MM1268, 73
MM5321, 30
MN67603NS, 35
MN67621F, 35
Monoscope, 101, 102, 114, 115, 116, 117, 118, 119, 120, 121, 122, 154
Monotube cameras, 160
MPEG IMX, 182
Multi Media Card (MMC), 183
Multiburst Generator, 110, 111, 112, 116

Multiple chip SPGs, 35
Multiple screen displays, 153
Multiplier Orthicons, 158

N

Nanni Loj, 173
Narragansett Imaging, 161
ND Filter, 171
Nematic molecules, 145
Neutral Density Filter, 171
Newton, Isaac, 52
Newvicon, 160
Nipkow disk, 8, 16, 134, 156
Nipkow, Paul Julius Gottlieb, 15, 16
Noise Level, 165
Non-standard SPGs, 39
Notch filter, 78
NTE7049, 33
NTE879, 73
NTSC & PAL Decoding, 76
NTSC & PAL Encoding, 52
NTSC (National Television System Committee), 19, 20, 21, 22, 23, 27, 33, 35, 40, 52, 61, 121, 130, 155, 162, 186

O

Objective Lens, 170
Octodes, 125
Odd Field, 26
OEL (Organic Electro Luminescence), 147
Optical Lens, 170
Organic Light Emitting Diode (OLED) Display, 147, 148, 154, 155
Orthicons, 158, 163
Oscilloscope, 126, 127, 129
Output Switch, 171
Overscan Area, OS Area, 139

P

P22, 136
P4, 136
PAL (Phase Alternation by Line), 17, 19, 20, 21, 23, 35, 52, 56, 57, 59, 60, 61, 78, 110, 118, 155, 191, 193
PAL delay line, 79
PAL Identification Circuit, 57
PAL Switching, 57
PAL+, 64

Index

PAL-D Decoders, 57
PAL-S Decoders, 57
Pantelegraph, 15
Pasecon, 160
Passive Pixel Sensor (PPS), 169, 170
Paycheck, John Woo's, 155
PDA (Personal Digital Assistant), 173
Pentodes, 125
Peritel, 96
Persistence of vision, 9, 25, 47
Perskyi, Constantin, 16
Phonovisor, 175
Phosphor, 136, 140
Photoelectric mosaic, 157, 158
Pickup Devices, 156
Pickup tube features, 163
Picture Enancher, 190
Picture In Picture, PIP, 94
Picture Phone, 18
Pincushion, 139
Pixel, 7, 165
Plasma Display Panel (PDP), 146
Plate, 125
PLATO, 146
Playstation3, 183
Plumbicon, 160, 161, 163, 164, 173
PM5440, PM5544, PM5644, 116, 119
Pope John XXIII, 173
Pre & Post Equalisation Pulses, 24
Precision In-Line Shadow Mask, 141
Primary Colours, 53
Princess Diana, 199
Pupil, 47
Purity, 139
Pyroelectric Vidicon, 161

Q
QL, 68
Quadrature Amplitude Modulation (QAM), 56
Quadrature Phase, 55
Quadruplex, 176, 177, 198, 199

R
Radiovision, 134
Raster, 25

Rasterizer, 131, 133
RCA (Radio Corporation of America), 17
Reinitzer, Friedrich Richard, 144
Resistron, 160
Retina, 46
Retrofocus Lens, 170
RGB colour changer, 113
RGB Colours, 53
RGB Components, 52
RGB Decoder, 82, 86, 88, 89, 98
RGB Encoder, 62
Rhodopsin, 47
Rods & cones, 47, 52

S
S178A, 35
SAA1043, 35
SAA1044, 35
SAA1101, 35
SAA7114/SAA7115/SAA7118, 98
Safe Area, 138
Sandcastle Pulse, 86
Sangster, Frederik Leonard Johan, 165
Sarnoff, David, 49
Saticon, 160, 161, 163
SC6433, 35
Scanning, 8, 9, 16, 21, 24, 28, 42, 45, 114, 127, 134, 142, 143, 157, 160
SCART, 64, 86, 91, 96, 97, 113, 121
Scene Files, 172
Schroeder, Alfred, 50
S-Correction, 139
Screen Grid, 125
SDA9257, 44
SDI (Serial Digital Interface), 189
SECAM (Systeme en Couleur a Memoire), 19, 20, 57, 59, 60, 62, 80, 81
SECAM decoders, 91
SECAM Decoding, 79
SECAM fire or flame, 60, 63
SECAM-IV, 64
Secure Digital (SD), 183
Seimart, 20
Sensitivity, 164
Set Top Box, 191
Shadow Mask, 140
ShibaSoku, 115

Index

Siemens, Ernst Werner von, 15
Signal Plate, 157
Silicon Diode Array, 160
Silicon Video Corporation, 143
Silicon-Vidicon, 160
Silvatone, 175
Single chip SPGs, 30
Single Line Selector, 42
Slottow, H. Gene, 146
Smith, George, 165
Smith, Oberlin, 174
Society of Motion Picture and Television Engineers (SMTPE), 104
Spectracal, 119
SPG625, 35, 38, 67, 70, 76, 101, 112
Spiral Television, 44, 45
Sputnik, 10
Spycam, 173
Spypen, 173
Standard TV, 21
Standards Converters, 21
Static Convergence, 140
STN=Super Twist Nematic, 145
Stripe Filter, 163
STV2180, 82
STV224XH/228XH/223XH, 91
STV9306, 91
Subtractive Mixing, 54
Super Betamax, 180
Super VHS, S-VHS, 58, 97, 180
Super8, 12
Suppressor Grid, 125
Surface-conduction Electron-emitter Display (SED), 144, 145
S-Video, 58, 72, 86, 94, 97, 98, 122, 172, 191, 193, 195, 196
Swinging Burst, 57
Swiss Federal Institute of Technology, 150
Sync extraction, 40
Sync Pulse Generators (SPGs), 29, 137
Sync Separator, 23
Synchronisms (Syncs), 23
Composite Synchronisms, 23
Synchronous Demodulation, 76
Synthetic Aperture, 121

Index

T

Tang, Chin, 147
Taynton, William Edward, 8
TBA520, 65
TBA520, TBA560 & TBA540, 81
TBA990, 81
TCA240, 65
TCA640 & TCA650, 81
TDA2140, TDA2151 & TDA216, 81
TDA2506 & TDA2507, 76
TDA2579, 44
TDA2595, 44
TDA3562A, 82
TDA3565, 82
TDA4555 & TDA4556, 81
TDA4660, 82
TDA8174A, 91
TDA8181, 44
TDA8362, TDA8366, 89
TDA8390, TDA8461, TDA8451, 90
TDA8395, 89
TDA8490, 90
TDA8501, 75
TDA8505, 75
TDA9181, 86, 89, 98
TDA935X/6X/8X, 91
TEA1002, 65
TEA2000, 65
Teer, Kees, 165
Telechrome, 142
Telecine, 175
Telectroscope, 15
Telegraphone, 174
Telephony, 8
Television, 8, 9, 10, 15, 21, 47, 156, 172
Televisor, 16, 134
Tension Mask, 141
Tesla, Nikola, 125
Test Card Maker (TCM), 120
Test cards and monoscopes, 114
Test Cards, 115
Test Pattern Maker, 121
Test Patterns, 101
Testbeeld, 118

Index

Testpattern Generator, 118
Tetrodes, 125
TFT (Thin Film Transistor), 145
The Baird Company, 101
Thick-Film Dielectric Electroluminescent (TDEL) Display, 148
ThinCRT, 143
Three-Tube CRT Projectors, 151
Time Base Corrector, 189
Time Code Selection, 172
Time Division Multiplexing (TDM), 64
Time Shifting, 179
Timebase Generator, 23
Timecode, 184
Timeslip, 183
Transverse Recording System, 176
Trichromacy, 47
Trichromaticism, 47
Trigger, 128
Trinicon, 160
Trinitron, 140, 141
Triode, 125
Triple Head, 153
Tripler, 137
TRT-1A, 176
Tube-based camera devices, 156
TVP5146, 91
TW9919, 98
Type 'A', 177
Type 'B', 178
Type 'C', 178

U

Ultra High Performance lamp, 152
Ultra High Pressure (UHP), 152
U-matic, 178, 183, 189
Unitron, 143
University of Illinois at Urbana-Champaign, 146
Unvarying, 55

V

V7040, 73
Vacuum Tube Diode, 125
Valdemar Poulson, 174
Valensi, Georges, 18
Varying, 55

Index

VBLK, 26
VCC, Video Compact Cassette, 180
VCR (Video Cassette Recorder), 178
Vectorscope, 133, 134
VERA (Vision Electronic Recording Apparatus), 176
Vertical & horizontal lines pattern, 108
Vertical Blanking, 26
Vertical Deflection Eletrodes, 128
Vertical Frequency, 26
Vertical Smear, 167
Vertical Sync Pulse, 24
VGA to PAL/NTSC Scan Converter, 121
VHS (Video Home System), 179, 183
VHS-C Recorder, 179, 180
VIC20, 58
Video 2000, V2000, 180, 181, 184, 185
Video 8, 13, 181
Video Bandwidth, 28
Video Cameras, 156
Video cassette recorder mechanisms, 183
Video conversion, 96
Video Heads, 177, 180, 184, 185, 186, 187
Video Hi8, (High-Band Video8), 181
Video Mini Cassette, 180
Video Projector, 50, 149, 151, 192
Video Recording Systems, 174
Video, 11, 12, 14
Videowall, 153
Vidicon, 160, 163
Virtual Retinal Display (VRD), 150
Vistacon, 160
Visual cortex, 47
Volta, Alessandro, 15, 125
VR-1000, 176
VS, 24
VTR Save/Standby Switch, 171

W

W3XK, 17
Warm White Colour, 171
Waveform Monitor, 130
Webcam, 173
White Balance Adjustment, 171
White Clip Circuit, 171
White Clip Level, 24

221

Index

White Peak Level, 24, 130, 188, 192
Williams, Richard, 145
Willson, Robert, 146
WOWvx Display, 149
Wright brothers, 15

X
X-Axis, 127
XEL-1, 148
X-Plates, 128
XY Electrostatic Deflection Yokes, 128

Y
Y/C 7-pin Connector, 58
Y-Amplifier Stage, 128
Yoshinaka, Kazuo, 150
Y-Plates, 128

Z
Z80, 173
Z-Axis, 128
Zebra Stripe Level Indicator, 171
Zenith Radio Corporation, 19
Zilog, 173
ZNA134, 30, 31
ZNA234, 39, 102
Zoom Lens, 170
Zworykin, Vladimir, 16, 17, 135, 157
ZX Spectrum, 30, 66
ZX128, 68
ZX81, 30
ZXFV4583, 44